AF411352

Electrical Properties of Model Lipid Membranes

Electrical Properties of Model Lipid Membranes

Editor

Monika Naumowicz

MDPI • Basel • Beijing • Wuhan • Barcelona • Belgrade • Manchester • Tokyo • Cluj • Tianjin

Editor
Monika Naumowicz
Faculty of Chemistry
University of Bialystok
Bialystok
Poland

Editorial Office
MDPI
St. Alban-Anlage 66
4052 Basel, Switzerland

This is a reprint of articles from the Special Issue published online in the open access journal *Membranes* (ISSN 2077-0375) (available at: www.mdpi.com/journal/membranes/special_issues/EP_MLM).

For citation purposes, cite each article independently as indicated on the article page online and as indicated below:

LastName, A.A.; LastName, B.B.; LastName, C.C. Article Title. *Journal Name* **Year**, *Volume Number*, Page Range.

ISBN 978-3-0365-4058-0 (Hbk)
ISBN 978-3-0365-4057-3 (PDF)

Contents

Preface to "Electrical Properties of Model Lipid Membranes"

This book presents the significant recent progress that has been made recently in the study of the electrical properties of various model membrane systems used for mimicking biological membranes. The book contains nine articles: eight research articles and one review. The complete description of each study and the main results are presented in more detail in the full manuscript, which the reader is invited to read.

Studies of the electrical properties of model lipid membranes have been carried out for many years. However, there are still many issues that have not been verified experimentally and for which the existing results are incomplete or inconsistent. Therefore, the main objective of this book was to collect recent scientific and review articles on the electrical properties of model lipid membranes. This objective has been successfully achieved, for which I express heartfelt appreciation to all authors and reviewers for their excellent contributions.

Monika Naumowicz
Editor

 *membranes*

Article

Electrophoretic Light Scattering and Electrochemical Impedance Spectroscopy Studies of Lipid Bilayers Modified by Cinnamic Acid and Its Hydroxyl Derivatives

Monika Naumowicz [1,*], **Marcin Zając** [2], **Magdalena Kusaczuk** [3], **Miroslav Gál** [4] **and Joanna Kotyńska** [1]

1. Department of Physical Chemistry, Faculty of Chemistry, University of Bialystok, K. Ciolkowskiego 1K, 15-245 Bialystok, Poland; joannak@uwb.edu.pl
2. Doctoral School of Exact and Natural Sciences, University of Bialystok, K. Ciolkowskiego 1K, 15-245 Bialystok, Poland; m.zajac@uwb.edu.pl
3. Department of Pharmaceutical Biochemistry, Medical University of Bialystok, Mickiewicza 2A, 15-222 Bialystok, Poland; mkusaczuk@wp.pl
4. Department of Inorganic Technology, Faculty of Chemical and Food Technology, Slovak University of Technology, Radlinského 9, 812 37 Bratislava, Slovakia; miroslav.gal@stuba.sk
* Correspondence: monikan@uwb.edu.pl; Tel.: +48-8573-880-71

Received: 23 October 2020; Accepted: 13 November 2020; Published: 15 November 2020

Abstract: Pharmacological efficiency of active compounds is largely determined by their membrane permeability. Thus, identification of drug-membrane interactions seems to be a crucial element determining drug-like properties of chemical agents. Yet, knowledge of this issue is still lacking. Since chemoprevention based on natural compounds such as cinnamic acid (CinA), *p*-coumaric acid (*p*-CoA) and ferulic (FA) is becoming a strong trend in modern oncopharmacology, determination of physicochemical properties of these anticancer compounds is highly important. Here, electrophoretic light scattering and impedance spectroscopy were applied to study the effects of these phenolic acids on electrical properties of bilayers formed from 1,2-dioleoyl-sn-glycero-3-phosphocholine (DOPC), 1,2-diacyl-sn-glycero-3-phospho-l-serine (PS) or DOPC-PS mixture. After phenolic acid treatment, the negative charge of membranes increased in alkaline pH solutions, but not in acidic ones. The impedance data showed elevated values of both the electrical capacitance and the electrical resistance. We concluded that at acidic pH all tested compounds were able to solubilize into the membrane and permeate it. At neutral and alkaline pH, the CinA could be partially inserted into the bilayers, whereas *p*-CoA and FA could be anchored at the bilayer surface. Our results indicate that the electrochemical methods might be crucial for predicting pharmacological activity and bioavailability of phenolic acids.

Keywords: cinnamic acid; *p*-coumaric acid; ferulic acid; phenolic compound; electrophoretic light scattering; electrochemical impedance spectroscopy; phospholipid bilayers; liposomes; drug-membrane interaction; membrane biophysical study

---

1. Introduction

A growing body of evidence indicates dietary polyphenols to be efficient antioxidants counteracting negative effects of oxidative stress accompanying many diseases such as cardiovascular diseases, inflammation, and also cancer [1–6]. In line with previous reports, many of these compounds may act as chemopreventive or even chemotherapeutic agents [5,7–9]. Specifically to malignant cells, cinnamic acid (CinA) and its hydroxy derivatives have been shown to exhibit both antioxidant a antineoplastic activity, with their cytostatic effect being largely related to their structural characteristics [10–13]. However, the mode of action of CinA and its derivatives is still incompletely determined, but may involve scavenging of reactive oxygen species, modulation of gene expression, activation of xenobiotics metabolism-related enzymes, and regulation of signal transduction pathways essential for tumor cell growth and progression [14].

Despite their natural origin and many promising biological effects, bioavailability of hydroxycinnamic acids presents certain limitations. Although working well in aqueous media, their hydrophilic nature is usually a restriction for lipophilic system protection [15]. Lipophilicity is a major physicochemical property of chemical substances, which affects their biological activities. It is known to be important for describing both pharmacodynamic and pharmacokinetic aspects of drug action. In biological systems, lipophilicity largely determines key properties of potential pharmacological agents, such as solubility of substances in biological fluids, penetration through the biological membranes, rate of absorption, affinity to plasma and tissue proteins, and distribution into the specific body compartments [16]. Lipophilicity is commonly expressed by the logarithm of n-octanol/water partition coefficient ($\log P$) for ionizable compounds of a neutral form of compounds. Initially, $\log P$ was considered important in drug and pesticide discovery and design, but now it is an essential characteristic of all chemicals. This is because $\log P$ largely determines chemicals fate both inside a living organism and in the environment. $\mathrm{Log}P$ values are typically between -3 (very hydrophilic) and $+10$ (extremely hydrophobic) [17]. For ionizable forms of compounds, the distribution coefficient ($\log D$) at a specific pH is also often used. As opposed to $\log P$, which is only valid for a single electrical species, $\log D$ represents the pH-dependent mixture of all electrical species occurring at given pH [18].

Based on the observation that most medication drugs are relatively small and lipophilic molecules, Lipinski et al. formulated "the rule of five" (Ro5) [19]. This is a rule of thumb in determining if a pharmacologically/biologically active chemical compound, a candidate for a drug, would be potentially bioavailable via oral administration in humans [20]. According to Ro5, chemicals are less prone to adsorb on the cell membrane and more likely to permeate the bilayer when their calculated n-octanol/water partition coefficient ($c\log P$) $\geq$ 5, they have $\geq$10 H-bond acceptors, $\geq$5 H-bond donors, and their molecular weight (M_{WT}) exceeds 500 g/mol [19]. Because each threshold is a multiple of 5, the rule was called Ro5. Molecules not complying with more than one of these rules may have problems with bioavailability [21].

To obtain lipophilicity parameters of significant pharmacokinetic and pharmacodynamic relevance, partition coefficients between (mostly but not exclusively) artificial membranes and water have been tested recently [22]. Monolayers and bilayers have been used for several decades as models of biomembranes to study solute/biological membranes interactions. A variety of lipids and their mixtures may serve as components to prepare both types of these membrane models. Nevertheless, phospholipids tend to be the most widely exploited as they are easily reproducible and have well standardized systems [23]. These lipids are either negatively charged or zwitterionic (electrically neutral due to an equal number of positive and negative charges). Phospholipids are amphipathic and occur naturally in all living organisms as the major components of cell membranes. Most biological membranes are characterized by the asymmetrical distribution of phospholipids within the inner and outer leaflets. The inner leaflet of the bilayer mainly consists of negatively charged lipids, such as phosphatidylserine (PS), while electrically neutral lipids, such as phosphatidylcholine (PC) and phosphatidylethanolamine (PE), are mostly located in the outer layer [24]. At physiological pH, negative charge of the outer leaflet lipids is due to low values of acid dissociation constant (pK_a) of the phosphate moieties of the lipid head group [25].

Our understanding of the properties, function and structure of biological membranes has benefitted a lot from the experimental and theoretical studies of the electrical properties of lipid bilayers. Therefore, evaluation of the parameters such as surface charge density (σ) of the membrane is critical for determination of the electrostatic interactions between membranes and their surrounding solutes. The surface charge of a lipid membrane, which may change with pH, depends also on the lipids present in the outer layer and can be quantified by zeta-potential measurements using electrochemical light scattering (ELS) technique. Zeta potential depends on a lots of parameters, including temperature, pH, conductivity (ionic strength), and solvent (viscosity). Small changes in any of these parameters can potentially dramatically affect the zeta potential values [26].

In order to best mimic electrochemical conditions observed in natural membranes, model systems of artificial bilayers ought to possess similar values of capacitance (C_m) and resistance (R_m) to those observed in biological membranes. In this respect, the parallel plate capacitor model may serve as an estimator of the lipid bilayer capacitance. Assuming that typical hydrophobic thickness of a membrane equals 4 nm and a relative permittivity of ranges from 2 to 4, the capacitance of the bilayer is supposed to be 0.5–1.0 μF/cm^2. In fact, C_m values placing within this range have been already confirmed in experimental research [27–29]. Typical values of bilayer resistance are in the range 10^4–10^7 Ω cm^2 [30,31]. The measurement of lipid membrane resistance may sometimes be difficult and unreproducible. These variations are most likely a result of a leakage at the bilayer support and/or trapped emulsified droplets [32]. Luckily, membrane capacitance and resistance can be easily measured by electrochemical impedance spectroscopy (EIS). EIS is a technique that enables to extract the information not only about the bulk phase of tested material (e.g., dielectric constant and conductivity) but also about their inner and outer interfaces (e.g., interfacial region capacitance and derived quantities) [33].

Herein, we report on the modulation of the electrical properties of model cell membranes by cinnamic acid and its two naturally occurring hydroxy derivatives: *p*-coumaric acid (*p*-CoA) and ferulic acid (FA). The selected phenolic acids are widely distributed in plants, fungi and algae and recognized as privileged structures for the development of bioactive compounds with therapeutic potential. These compounds are structurally related (Figure 1) and a correlation between their structures and behavior in the surrounding solution seems to warrant further investigations. Although a considerable amount of data concerning hydroxycinnamic acids interactions with membranes exists, still little is known about the electrical properties of bilayer lipid membranes modified by CinA, *p*-CoA or FA. Therefore, this paper is focused on the effect of cinnamic acid and its derivatives on the resistance, capacitance, and the surface charge density of model membranes (spherical lipid bilayers and liposomes).

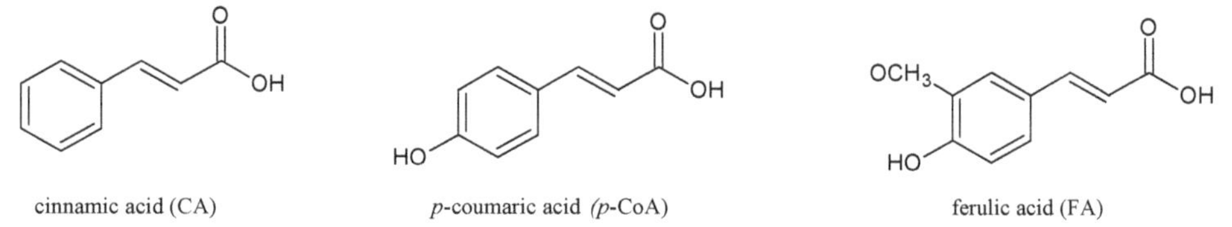

cinnamic acid (CA)   *p*-coumaric acid *(p*-CoA)   ferulic acid (FA)

Figure 1. Chemical structures of compounds under study.

We previously described changes in the electrical parameters caused by e.g., membrane compositions in lipid–lipid [34], lipid–sterol [35], lipid–fatty acid [36], or lipid-carotenoid membranes [37]. The investigations reported in this paper are a continuation of the studies on the interaction of model membranes of increasing complexity with naturally occurring phenolic compounds. The bilayers were formed from 1,2-dioleoyl-sn-glycero-3-phosphocholine (DOPC), 1,2-diacyl-sn-glycero-3-phospho-l-serine (PS) or from DOPC-PS mixture in concentrations of 9:1 and 8:2 mol%, respectively, which corresponds to PS content in the human cerebral cortex [38]. Next, phospholipid membranes were modified by CinA, *p*-CoA or FA in the concentrations determined on the basis of MTT analysis shown in previous studies on human glioblastoma cell lines. Afterwards, we analyzed the influence of the pH of the electrolyte solution and the composition of membranes on their surface charge to better describe interactions in model membranes modified by the phenolic acids. We also analyzed if CinA, *p*-CoA and FA were capable of altering the electrical resistance and capacitance of the bilayers. The results of our research should give useful indications in understanding the role of chemical structure of hydroxycinnamic acids in determining their interactions with the microenvironment of model lipid bilayer and, thereby, allow us to speculate about the in vivo bioavailability of the investigated compounds.

2. Materials and Methods

2.1. Reagents

Compounds: trans-cinnamic acid (≥98.0%), *p*-coumaric acid (≥98%), trans-ferulic acid (99%), 1,2-dioleoyl-sn-glycero-3-phosphocholine (>99%) and 1,2-diacyl-sn-glycero-3-phospho-l-serine (≥97%) were obtained from Sigma-Aldrich (St. Louis, MO, USA).

The remaining reagents had the highest degree of purity available on the market and only freshly prepared solutions were used. The electrolyte solutions (155 mM/L NaCl) were prepared using deionized water purified to a resistance of 18.2 MΩ (HLP 5UV System, Hydrolab, Hach Company, Loveland, CO, USA) and filtered using a 0.2-μm membrane filter to eliminate any impurities. All glassware and equipment were cleaned with 18.2 MΩ cm of ultrapure water.

All experiments were conducted at a mean room temperature of 20 ± 2 °C.

2.2. Methods

2.2.1. Preparation of Liposomes

Liposomes were prepared by the thin film hydration method to obtain small unilamellar vesicles (SUVs) [39]. Briefly, DOPC, PS, CinA, *p*-CoA or FA were dissolved in chloroform (anhydrous, ≥99%). Lipid alone and phenolic acids solutions were transferred into a glass tube in amounts suitable to obtain the appropriate concentrations of phenolic acids. The concentrations were chosen on the basis of the results of MTT assay performed on human glioblastoma cell lines subjected to CinA, *p*-CoA or FA treatment for 24 and 48 h. Regarding the IC_{50} values and in order to choose the sublethal doses of phenolic acids, the concentrations of 1.0 and 5.0 mmol/dm^3 have been selected [10,40,41]. A thin lipid film was obtained after the organic solvent evaporation under an argon gas stream to remove any organic solvent residue. The dried lipid film was hydrated with an electrolyte solution (155 mM/L NaCl). Liposomes were formed by sonicating the suspension using the ultrasound generator (UD 20, Techpan, Warsaw, Poland). Sonication was repeated five times, 90 s each. Since heat is released during the process, cooling the suspension is mandatory. It was performed with an ice bath (a container with a mixture of ice and dry sodium chloride). The samples consisted of plain liposomes (10 mg of: DOPC, PS, 9:1 mol% DOPC:PS or 8:2 mol% DOPC:PS), as well as liposomes containing chosen phenolic acid. The liposomes were directly examined in the ELS apparatus.

2.2.2. Preparation of Spherical Bilayers

The stock solutions for bilayers formation were composed of 20 mg cm^{-3} of DOPC or DOPC-PS mixture in concentrations of 9:1 and 8:2 mol%, respectively. Chloroform phospholipids alone and phenolic acids solutions were transferred into a glass tube in amounts suitable to obtain the same concentration of CinA, *p*-CoA or FA as in liposome forming solutions. After mixing the compounds, the solvent was removed under a nitrogen flow, and the resulting dry residues were dissolved in a n-hexadecane-n-butanol mixture (10:1 by volume). All the bilayer-forming solutions were stored in darkness at a temperature of <4 °C in a refrigerator for at least 3 days before examination. To minimize the oxidation of lipids, the vessels containing final samples were filled with nitrogen.

Bilayers were prepared by the method of squeezing the solution, which enables creating spherical bilayers dividing two aqueous solutions. They were formed at the Teflon cap constituting a part of the measuring vessel. During membranes creation, the solvent mixture was removed from the lipid phase, resulting in membranes with the same composition as in stock solutions. Bilayers were monitored both electrically and optically during the entire process of formation. Measurements were initiated 10–15 min after the membranes turned completely black. Membrane images were taken by color CCD camera using the WinFast PVR program. The bilayer areas were calculated from the photographs, taking into consideration the spherical nature of the surface and employing the Makroaufmassprogram program [42]. The area of the spherical membranes was about 6×10^{-2} cm^2. More details regarding the procedure for the membrane formation are given in our previous works [31,36,37].

2.2.3. Electrophoretic Light Scattering Measurements

Microelectrophoretic mobility of the liposomes was obtained by performing micro-electrophoretic assessments on samples using ELS technique. The measurements were performed by Zetasizer Nano ZS apparatus (Malvern Instruments, Malvern, United Kingdom). The xperiment was carried out as a function of pH using a WTW InoLab pH 720 laboratory meter (WTW, Weinheim, Germany). The samples were suspended in 155 mM/L NaCl solution and titrated to the desired pH (range 2–10, every ± 0.3 units) with HCl or NaOH. Six measurements were made (each covering 100–200 series for a duration of 5 s) for each pH value per sample. Experiments were performed three times with similar results.

The surface charge density δ depends on electrophoretic mobility as described by the following equation [43]:

$$\delta = \frac{\eta \cdot u}{d} \tag{1}$$

where η is the viscosity of the solution, u is the electrophoretic mobility, and d is called the diffuse layer thickness.

The diffuse layer thickness is defined as [44]:

$$d = \sqrt{\frac{\varepsilon \cdot \varepsilon_0 \cdot R \cdot T}{2 \cdot F^2 \cdot I}} \tag{2}$$

where: R represents the gas constant, T—the temperature, F—the Faraday constant, I—the ionic strength of 0.9% NaCl, and ε and ε_0 refer to the permeability of the electric medium.

2.2.4. Liposome Size Determination

Dynamic light scattering (DLS) method was used to determine the size of the liposomes. The diameter of the particles was evaluated from the intensity of the dispersed light, which is the standard parameter measured, by the Zetasizer Nano ZS software. A helium–neon (He–Ne) ion laser at 633 nm wave length was used as the incident beam.

2.2.5. Electrochemical Impedance Measurements

Electrochemical impedance spectroscopy was conducted with an Autolab potentiostat (Model PGSTAT302N, Metrohm, Poland) coupled with an FRA2 module. A four-probe cell with two Ag/AgCl reference electrodes was employed for sensing and two platinum electrodes were used to carry the current for the measurements. The cell was presented in [37]. EIS data were registered in the frequency range of 0.1–100,000 Hz. A sinusoidal voltage excitation of 4 mV versus open circuit potential was applied. No direct current was used.

Analysis of the data obtained during EIS measurements was performed using NOVA software (v. 1.10) with the fitting model shown in Scheme 1. This electrical circuit was described previously for the lipid bilayer unmodified by channels formers or carriers [45]. The circuit consists of the solution resistance (R_e) in series with a membrane resistance (R_m) and a membrane capacitance (C_m), which are in parallel and correspond with the electric and dielectric properties of the bilayer. The values of the R_m and C_m were normalized to the bilayer area. Each value represents the average of 6–10 independent measurements.

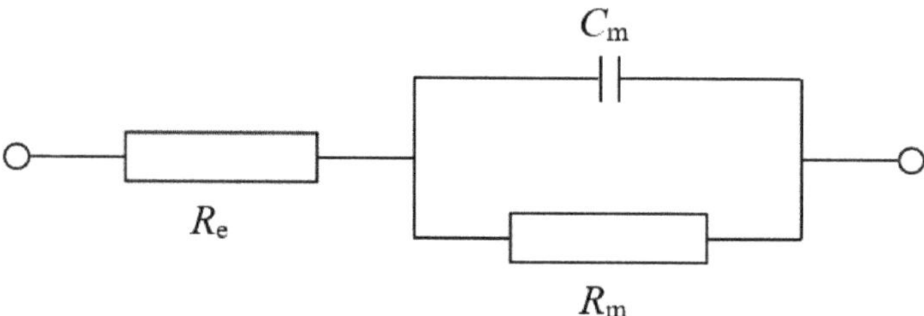

Scheme 1. The equivalent circuit model used to fit EIS data.

3. Results

In order to get a deeper insight into the interactions between CinA, *p*-CoA or FA and bilayer lipid membranes, a systematic series of experiments was carried out using two representative concentrations of each phenolic acid. From the results of previous tests carried out on human glioblastoma cell lines [10,40,41], 1 and 5 mM/L concentrations have been selected. Firstly, the surface charge density of phospholipid liposomal membranes (DOPC, PS, DOPC/PS 9:1 and DOPC/PS 8:2), plain and modified by phenolic acids, was calculated based on the electrophoretic light scattering (ELS) technique. Subsequently, the capacitance and resistance values of DOPC, DOPC/PS 9:1 or DOPC/PS 8:2 spherical bilayers (plain and phenolic acid-modified) were obtained on the basis of the impedance data.

3.1. The Effect of Cinnamic Acid and Its Derivatives on Surface Charge Densities of Liposomal Membranes

To get insight into the surface charge density-modulatory properties of CinA, *p*-CoA and FA, the ELS technique has been engaged. The measurements were performed as a function of H^+ concentration. In order to receive the values of pH-dependent electrophoretic mobility, liposomes were suspended in 155 mM/L NaCl solution, which was further titrated to the adequate pH with concentrated NaOH or HCl. Surface charge density values were calculated on the basis of electrophoretic mobility in accordance to Equation (1) provided in section "Materials and Methods". Representative plots from at least three independent experiments are shown.

Surface charge density dependence on pH of plain and CinA-modified DOPC, DOPC/PS 9:1, DOPC/PS 8:2 and PS liposomal membrane occurred to be similarly shaped (Figure 2). An increase in positive surface charge density together with the decrease in pH values was noticed, however, only to a certain point. Conversely, along with the increasing pH, the negative charge of the membranes incremented until the plateau was reached. The visible decrease in the surface charge of PC-containing membranes at pH ~ 8.0 was possibly caused by the destruction of the membrane structure at such a high pH, which is in line with earlier reported data [7]. Our results demonstrated that in low pH, any visible alterations in values of the surface charge density were detected for all tested liposomal membranes treated with CinA. Nevertheless, at high pH values, the presence of CinA resulted in visible changes in surface charge densities of the aforementioned membranes (the negative charge of the membranes increased) compared to data obtained for plain phospholipid liposomal membranes. Those changes were apparently dose-dependent.

The liposomes treated with *p*-coumaric acid were examined analogously (Figure 3). It was found that in acidic solution, the presence of *p*-CoA did not affect the σ values. However, in alkaline solutions, an increase in the negative charge of *p*-CoA-treated membranes in comparison to plain phospholipid liposomal membranes was observed.

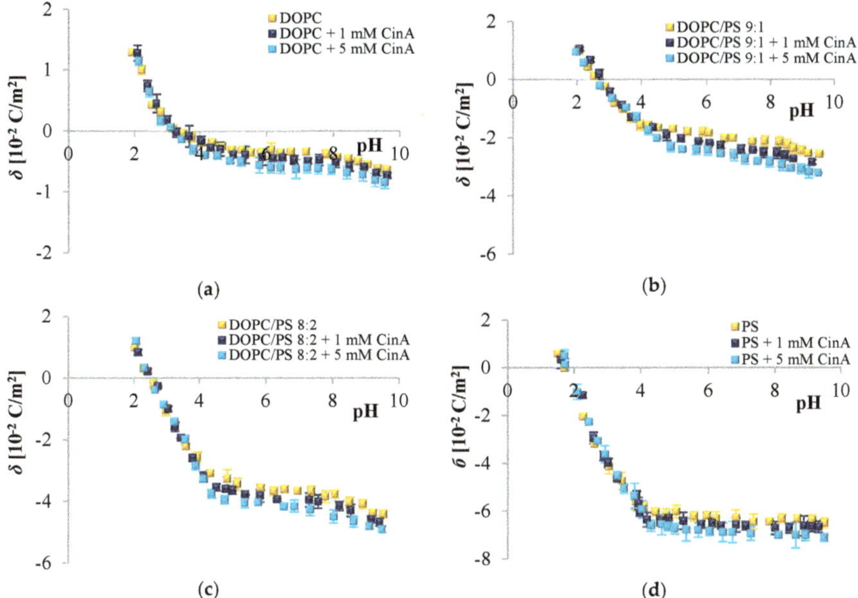

Figure 2. pH dependence of surface charge density of (**a**) DOPC, (**b**) DOPC/PS 9:1, (**c**) DOPC/PS 8:2, and (**d**) PS liposomal membranes as a function of (■) 0, (■) 1, and (■) 5 mM/L of cinnamic acid concentration. The results represent mean values from three independent experiments run in triplicate.

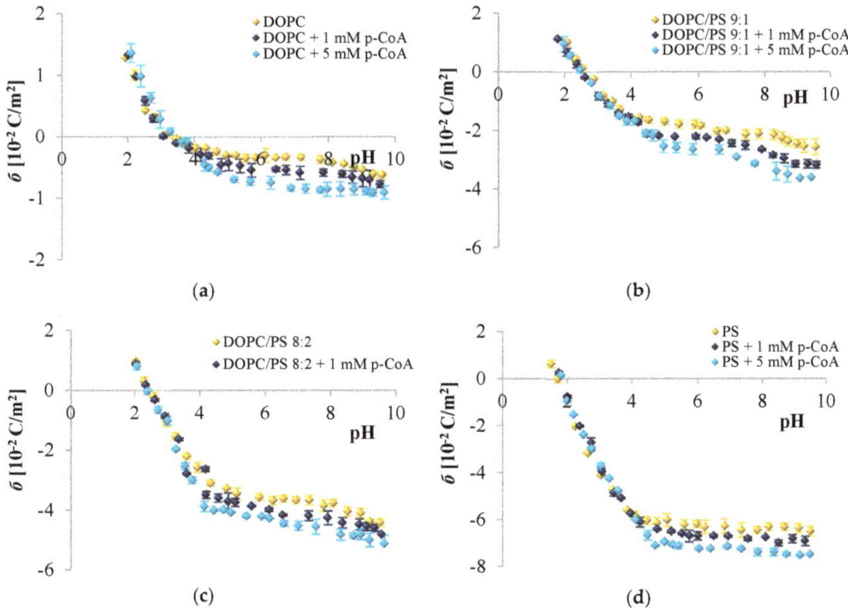

Figure 3. pH dependence of surface charge density of (**a**) DOPC, (**b**) DOPC/PS 9:1, (**c**) DOPC/PS 8:2, and (**d**) PS liposomal membranes as a function of (♦) 0, (♦) 1, and (♦) and 5 mM/L of *p*-coumaric acid concentration. The results represent mean values from three independent experiments run in triplicate.

The surface charge densities of the DOPC, DOPC/PS 9:1, DOPC/PS 8:2, and PS liposomal membrane, plain and modified with ferulic acid, are shown as a function of pH in Figure 4. Again, the presence of the phenolic acid did not affect the σ values at low pH, but caused an increase in the negative σ values at higher pH ranges.

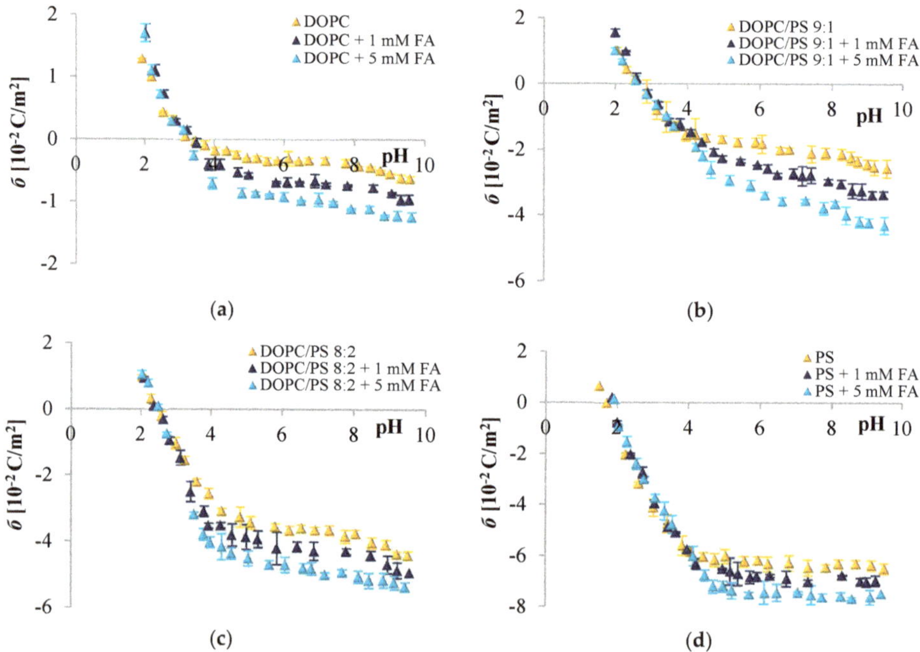

Figure 4. pH dependence of surface charge density of (**a**) DOPC, (**b**) DOPC/PS 9:1, (**c**) DOPC/PS 8:2, and (**d**) PS liposomal membranes as a function of (▲) 0, (▲) 1, and (▲) 5 mM/L of ferulic acid concentration. The results represent mean values from three independent experiments run in triplicate.

Considering the ELS data presented in Figures 2–4 and comparing membranes with the same phospholipid composition, it can be observed that the smallest changes in the surface charge density values were obtained in the presence of the most hydrophobic of all, cinnamic acid, whereas the most prominent alterations were noticed in the presence of the most hydrophilic compound, ferulic acid.

Finally, the results collected in Figures 2–4 illustrate that the presence of CinA, p-CoA or FA did not change the value of the isoelectric point of all analyzed phospholipid membranes. Since the isoelectric point of the membrane is one of the most important parameters describing its variable-charge surfaces, it is important to point out that it has shifted towards lower pH values for DOPC as compared to PS (decrease from pH ~3.5 to pH ~1.7, respectively). This is coherent with other investigations demonstrating that the isoelectric point of PC lipids fits into the pH values ranging between 3.2 and 4.0 [41,46,47], while reaching 1.4–1.7 values for PS lipids [41,46].

3.2. The Effect of Cinnamic Acid and Its Derivatives on the Size of the Phospholipid Liposomes

The effect of the analyzed phenolic acids on the size of the phospholipid liposome systems is presented in Tables 1–3. The modification of liposome membranes with CinA, FA and p-CoA resulted in variations in the size of the examined systems. Notably, the diameter of the liposomes was also found to be influenced by the applied doses of the phenolic acids (1 and 5 mM/L). These changes are observed in both acidic (pH ~ 2) as well as alkaline (pH ~ 9) pH.

Table 1. Effect of cinnamic acid on the diameter of the liposomes at different pH.

System	Diameter [nm]	
	pH ~ 2	pH ~ 9
DOPC	180.2 ± 3.2	205.5 ± 4.6
+1 mM/L CinA	195.2 ± 4.1	220.2 ± 3.6
+5 mM/L CinA	203.0 ± 2.0	234.2 ± 2.4
DOPC/PS 9:1	257.1 ± 4.2	219.2 ± 4.6
+1 mM/L CinA	253.0 ± 4.2	234.3 ± 3.6
+5 mM/L CinA	259.5 ± 3.0	248.0 ± 2.4
DOPC/PS 8:2	290.2 ± 2.2	240.2 ± 4.5
+1 mM/L CinA	297.0 ± 4.4	262.3 ± 4.6
+5 mM/L CinA	305.2 ± 2.1	277.4 ± 1.4
PS	340.3 ± 3.3	270.0 ± 2.1
+1 mM/L CinA	348.1 ± 4.5	284.1 ± 4.8
+5 mM/L CinA	360.3 ± 2.4	298.0 ± 3.1

Table 2. Effect of *p*-coumaric acid on the diameter of the liposomes at different pH.

System	Diameter [nm]	
	pH ~ 2	pH ~ 9
DOPC	180.2 ± 3.2	205.5 ± 4.6
+1 mM/L *p*-CoA	188.4 ± 3.1	227.3 ± 2.5
+5 mM/L *p*-CoA	200.1 ± 4.0	241.2 ± 2.8
DOPC/PS 9:1	257.1 ± 4.2	219.2 ± 4.6
+1 mM/L *p*-CoA	259.2 ± 2.3	240.4 ± 3.0
+5 mM/L *p*-CoA	270.1 ± 2.6	260.1 ± 2.9
DOPC/PS 8:2	290.2 ± 2.2	240.2 ± 4.5
+1 mM/L *p*-CoA	299.0 ± 5.4	273.0 ± 3.4
+5 mM/L *p*-CoA	310.5 ± 3.5	290.3 ± 2.4
PS	340.3 ± 3.3	270.0 ± 2.1
+1 mM/L *p*-CoA	351.1 ± 3.0	290.2 ± 4.0
+5 mM/L *p*-CoA	360.1 ± 2.7	315.4 ± 3.1

Table 3. Effect of ferulic acid on the diameter of the liposomes at different pH.

System	Diameter [nm]	
	pH ~ 2	pH ~ 9
DOPC	180.2 ± 3.2	205.5 ± 4.6
+1 mM/L FA	189.1 ± 4.0	249.2 ± 3.1
+5 mM/L FA	201.4 ± 4.5	265.2 ± 3.8
DOPC/PS 9:1	257.1 ± 4.2	219.2 ± 4.6
+1 mM/L FA	268.3 ± 5.3	248.6 ± 3.0
+5 mM/L FA	280.4 ± 5.6	260.4 ± 2.9
DOPC/PS 8:2	290.2 ± 2.2	240.2 ± 4.5
+1 mM/L FA	298.4 ± 3.4	276.5 ± 3.4
+5 mM/L FA	319.6 ± 3.5	301.7 ± 2.4
PS	340.3 ± 3.3	270.0 ± 2.1
+1 mM/L FA	347.5 ± 3.5	299.0 ± 4.4
+5 mM/L FA	359.1 ± 2.7	320.1 ± 3.4

Considering the data collected in Table 1, it can be observed that for plain DOPC liposomes as well as for DOPC liposomes treated with cinnamic acid, sizes of the liposomes increase along with increasing pH of the medium. The same tendency was observed for membranes treated with *p*-coumaric and ferulic acids (Tables 2 and 3). However, the opposite trend was noted for negatively charged liposomes (DOPC/PS 9:1, DOPC/PS 8:2, PS). Here, sizes of both plain and modified liposomes decreased together with increasing pH. In acidic pH, the diameter of phospholipid liposomes modified with cinnamic acid did not vary significantly from the diameter of liposomes modified with p-coumaric and ferulic acids. In alkaline pH, phospholipid liposomes treated with ferulic acid exhibited the largest sizes, whereas those treated with cinnamic acid were the smallest.

Figure 5 presents typical pH dependence of the diameter of the liposomes created with pure DOPC and DOPC modified with one of the analyzed phenolic acids, FA. Of note, in pH > 4, the sizes of plain DOPC liposomes and DOPC modified with ferulic acid increased with increasing pH of the medium, whereas in pH ~ 4 (isoelectric point of the system), the diameter of analyzed liposomes was the lowest.

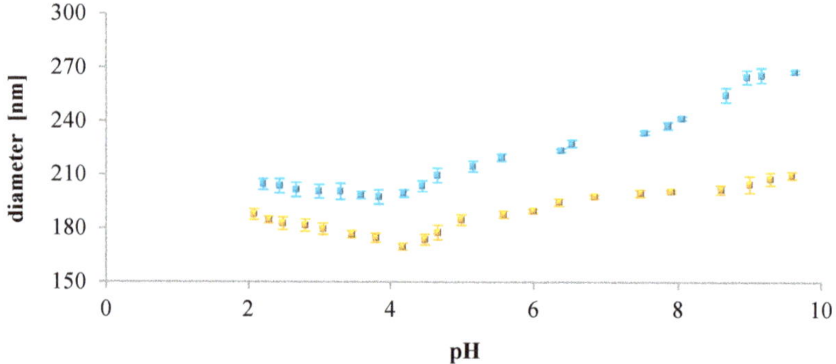

Figure 5. Typical pH dependence of the diameter of the liposomes formed by DOPC (■) and DOPC modified with 5 mM/L ferulic acid (■). The results represent mean values from three independent experiments run in triplicate.

3.3. The Effect of Cinnamic Acid and Its Derivatives on Capacitance and Resistance of Spherical Bilayers

Electrochemical impedance spectroscopy was utilized to quantify the effect of the examined phenolic acids on the membrane resistance (reciprocal of conductance) and capacitance. This method seems to be non-destructive and presents high sensitivity to feasible drug-induced modulation of bilayer thickness or packing [48]. As such, EIS was used over a wide range of frequencies from 10^{-1} Hz to 10^4 Hz under small amplitude. The 155 mM/L NaCl electrolyte solution was used to register impedance spectra.

Two different phospholipids i.e., DOPC and PS, were used to prepare one- or two-component bilayers. Unfortunately, bilayers composed only from PS did not have the ability to form sufficiently stable membranes for the impedance measurements. CinA, *p*-CoA or FA was introduced to the phospholipid model membrane solution at concentrations of 1 and 5 mM/L. In order to extract the R_m and C_m values from the impedance spectra, the EIS data were fitted to the equivalent circuit model shown in Scheme 1 presented in section "Materials and Methods". In the following analyses, the values of the impedance parameters referred to the bilayer surface area unit. At least six spherical lipid bilayers were tested for consistency. For quantitative data, average values with standard deviations are reported.

Figure 6 displays the impedance response in the form of Nyquist plots of different spherical bilayers, with varied concentrations of CinA. Notably, the plot registered for each bilayer contains a capacitive arc across the entire frequency range. The diameter of the arc corresponds to the membrane resistance. The incorporation of the CinA into the DOPC, DOPC/PS 9:1 or DOPC/PS 8:2 membrane resulted in an increase in the R_m values (compared to the values obtained for bilayers with the same lipid composition but without phenolic acid), confirming that this compound is active at the tested concentrations.

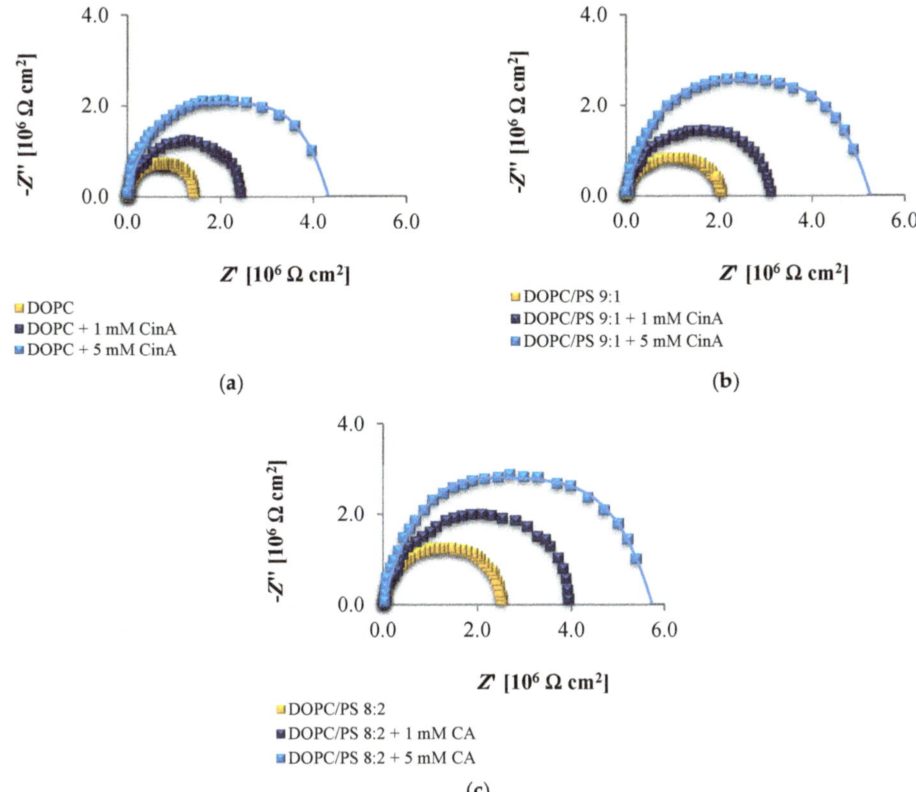

Figure 6. Representative impedance spectra for (**a**) DOPC, (**b**) DOPC/PS 9:1, and (**c**) DOPC/PS 8:2 spherical bilayers as a function of (■) 0, (■) 1, and (■) 5 mM/L of cinnamic acid concentration. Experimental data points are indicated with markers; model fits are shown with lines.

Table 4 provides the values of the electrochemical elements obtained by fitting the electrochemical impedance spectra. The typical value of the electrolyte resistance was equal to $(5.57 \pm 0.20) \times 10^3\ \Omega$ and was considered irrelevant to the bilayers characteristics. Therefore, this parameter was omitted in the curve fitting procedure. The R_m value of the DOPC bilayer was evaluated prior to the addition of the phenolic acid and found to be $(1.472 \pm 0.07) \times 10^6\ \Omega\ cm^2$, which is consistent with the reported values for other PC known to typically range between 10^4 and $10^7\ \Omega\ cm^2$ [30,31]. The obtained C_m value for plain DOPC membrane was $(0.617 \pm 0.03)\ \mu F/cm^2$, which is in line with the capacitance values of PC bilayers calculated based on the chronopotentiometric [27], chronoamperometric [49] or impedance [45] method. As presented in Table 4, the increase in membrane capacitance was also noticed after the addition of CinA to the spherical bilayers.

Three independent curves obtained for membranes with the same phospholipid composition, presented in Figure 7a–c, clearly show that adding increasing concentrations of *p*-CoA to the DOPC, DOPC/PS 9:1 or DOPC/PS 8:2 bilayers led to an increase in the membrane resistance.

Table 4. Effect of cinnamic acid on the resistance and capacitance of lipid bilayer.

System	Parameters	
	R_m [10^6 Ω cm^2]	C_m [10^{-6} µF/cm^2]
DOPC	1.472 ± 0.07	0.617 ± 0.03
+1 mM/L CinA	2.419 ± 0.10	0.637 ± 0.06
+5 mM/L CinA	4.094 ± 0.05	0.682 ± 0.04
DOPC/PS 9:1	2.043 ± 0.04	0.628 ± 0.03
+1 mM/L CinA	3.111 ± 0.10	0.644 ± 0.05
+5 mM/L CinA	4.836 ± 0.05	0.675 ± 0.03
DOPC/PS 8:2	2.521 ± 0.07	0.640 ± 0.05
+1 mM/L CinA	3.941 ± 0.10	0.651 ± 0.03
+5 mM/L CinA	5.777 ± 0.06	0.678 ± 0.02

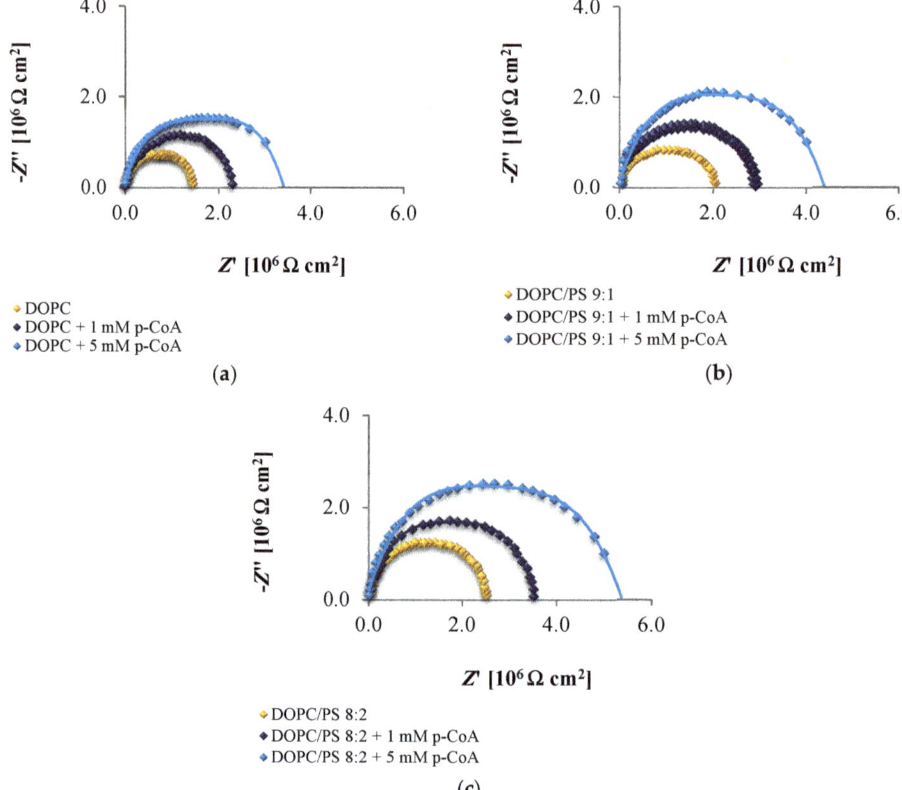

Figure 7. Representative impedance spectra for (**a**) DOPC, (**b**) DOPC/PS 9:1, and (**c**) DOPC/PS 8:2 spherical bilayers as a function of (♦) 0, (♦) 1, and (♦) 5 mM/L of *p*-coumaric acid concentration. Experimental data points are indicated with markers; model fits are shown with lines.

Table 5 summarizes the results concerning the influence of the membrane composition on the capacitance and resistance values. Upon introducing *p*-CoA, an increase in both parameters was detected compared to the plain membranes.

Table 5. Effect of *p*-coumaric acid on the resistance and capacitance of lipid bilayer.

System	Parameters	
	R_m [10^6 Ω cm^2]	C_m [10^{-6} µF/cm^2]
DOPC	1.472 ± 0.07	0.617 ± 0.03
+1 mM/L *p*-CoA	2.267 ± 0.12	0.645 ± 0.05
+5 mmol/dm^3 *p*-CoA	3.388 ± 0.08	0.743 ± 0.04
DOPC/PS 9:1	2.043 ± 0.04	0.628 ± 0.03
+1 mM/L *p*-CoA	2.877 ± 0.07	0.649 ± 0.02
+5 mM/L *p*-CoA	4.124 ± 0.04	0.733 ± 0.03
DOPC/PS 8:2	2.521 ± 0.07	0.640 ± 0.05
+1 mM/L *p*-CoA	3.505 ± 0.07	0.656 ± 0.04
+5 mM/L *p*-CoA	5.013 ± 0.04	0.743 ± 0.02

From the subsequent EIS data presented in Figure 8 and Table 6, it is evident that both the bilayer resistance and capacitance increase in the presence of FA.

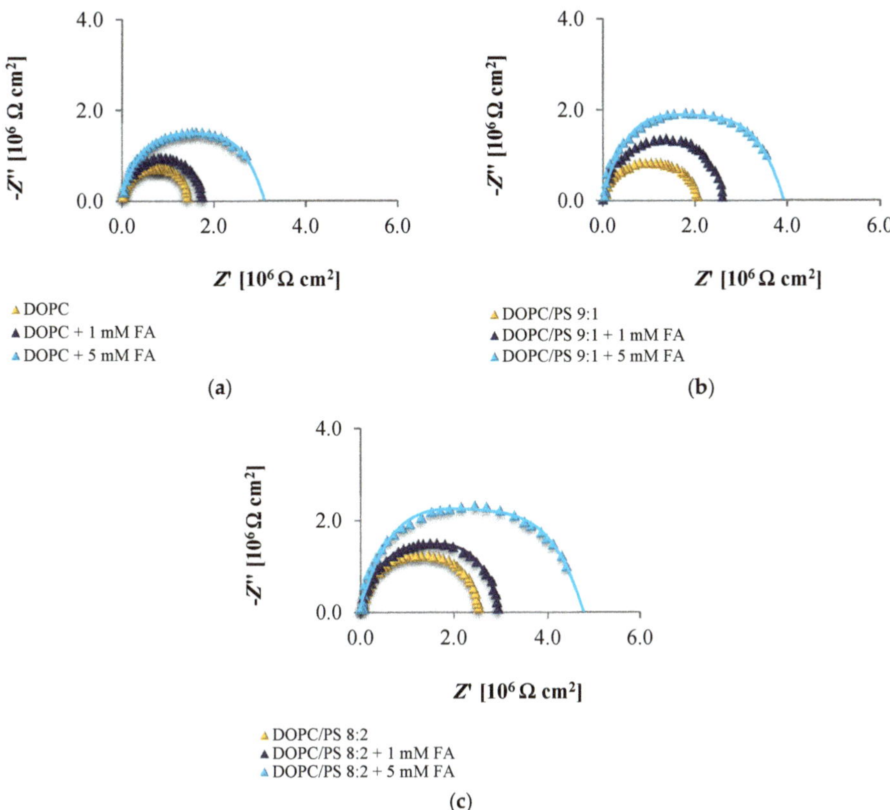

Figure 8. Representative impedance spectra for (**a**) DOPC, (**b**) DOPC/PS 9:1, and (**c**) DOPC/PS 8:2 spherical bilayers a function of (▲) 0, (▲) 1, and (▲) 5 mM/L of ferulic acid concentration. Experimental data points are indicated with markers; model fits are shown with lines.

Table 6. Effect of ferulic acid on the resistance and capacitance of lipid bilayer.

System	Parameters	
	R_m [10^6 Ω cm^2]	C_m [10^{-6} μF/cm^2]
DOPC	1.472 ± 0.07	0.617 ± 0.03
+1 mM/L FA	1.734 ± 0.04	0.651 ± 0.05
+5 mM/L FA	2.809 ± 0.08	0.755 ± 0.05
DOPC/PS 9:1	2.043 ± 0.04	0.628 ± 0.03
+1 mM/L FA	2.653 ± 0.07	0.651 ± 0.02
+5 mM/L FA	3.892 ± 0.04	0.743 ± 0.03
DOPC/PS 8:2	2.521 ± 0.07	0.640 ± 0.05
+1 mM/L FA	2.890 ± 0.07	0.657 ± 0.04
+5 mM/L FA	4.621 ± 0.04	0.748 ± 0.02

Taking into account the data gathered in Figures 6–8 and in Tables 4–6 and by comparing membranes with the same phospholipid composition, it may be noted that the smallest changes in the capacitance and the biggest changes in the resistance values were observed in the presence of the most hydrophobic of all tested compounds—cinnamic acid. Conversely, the largest alterations in the capacitance and the lowest in the resistance were noticed in the presence of the most hydrophilic one, ferulic acid.

4. Discussion

Cinnamic acid and its derivatives have attracted the attention of more and more researchers due to the broad variety of pharmaceutical and biological activities and due to the wide technological or industrial applications [50]. Polyphenols are important primarily because of their antioxidant property, antitumor activity or hypertension-preventing effect. Interactions of these phenolic acids with cell membranes play an essential role in their transport, distribution, selectivity, activity, and toxicity.

Small organic acids such as hydroxycinnamic acids have been postulated to interact with/and penetrate into biological membranes, however, their ability to demonstrate such an interaction with lipid membranes can be greatly influenced by the type of substituents attached to the main structure. Therefore, structurally similar compounds may differently interact with biological membranes, because of the complex relationship (which can be modulated by pH changes) between liposolubility and permeability [51]. The dissociation constant is a very important parameter describing solubility, lipophilicity, and permeability properties of compounds. Phenolic acids in aqueous solutions exhibit protonated and deprotonated forms. Analyzing the values of dissociation constants of the carboxyl and hydroxyl groups of the studied phenolic acids (Table 7), it can be stated that their molecular and monodeprotonated forms prevail within the range of physiological pH values. Moreover, the corresponding dissociation constant values are all similar to each other, which demonstrates that they are not a main factor determining the activity of the hydroxycinnamic acids. However, these data could be useful in studies concerning bioavailability and pharmacokinetics of potential pharmacological agents [52].

Table 7. Basic physicochemical parameters of cinnamic acid and its derivatives.

Compound	M_{WT} [1]	pK_{COOH} [2]	pK_{OH} [3]	$clogP$ [4]	$logD$ [5]	HBA [6]	HBD [7]
CinA	148.16 [18]	4.44 [53]	–	1.98 [54]	0.30 [18]	2 [55]	1 [55]
p-CoA	164.16 [18]	4.64 [56]	9.50 [56]	1.54 [54]	0.06 [18]	3 [55]	2 [55]
FA	194.18 [18]	4.46 [57]	8.75 [57]	1.42 [54]	−0.04 [18]	4 [55]	2 [55]

[1] molecular weight, [2] dissociation constant of the carboxyl groups, [3] dissociation constant of the hydroxyl groups, [4] calculated n-octanol/water partition coefficient, [5] distribution coefficient at pH 5.2 [6] hydrogen bond acceptor groups, [7] bond donor groups.

The positive values of the calculated log$P$ of CinA, p-CoA and FA may indicate that these acids are partitioning in the octanol phase. Nevertheless, only lipophilicity of uncharged molecules can be determined by this method, thus it is difficult to apply to these phenolic acids, which at physiological conditions are rather negatively charged. Analysis of Table 7 revealed that there are relevant differences between clog$P$ and log$D$ coefficient partitions for each acid. These findings suggest that the drug distribution between the aqueous solution and the lipid membrane is regulated by the ionization state of the molecule. All three compounds tested here are ionizable molecules, and it is already widely recognized that ionizable drugs partition into the lipid membrane to a high extent thanks to electrostatic interactions and formation of hydrogen bonds with polar groups of the phospholipids [18]. As a matter of fact, several reports inform about different experimental log$D$ values from the predicted log$P$ values calculated exclusively for the neutral species of drugs [58,59].

Rocher et al. reported the percentage of the undissociated forms of i.e., CinA, p-CoA and FA in pulvinar cells of *Mimosa pudica* L. at pH 5.2, which amounted to 70.4%, 58.6%, and 52.7% of the total pool of these acids respectively [18]. Considering the fact that only lipophilic neutral forms are able to cross the plasma membrane by diffusion, this parameter bears important information on this family of carboxylic acids [60].

Regarding the importance of physicochemical properties of drugs, Lipinski developed Ro5 [18]. According to this rule, candidates for efficient drugs should be characterized by log$P \leq 5$, molecular weight ≤ 500, number of hydrogen bond acceptors (O) ≤ 10, and number of hydrogen bond donors (OH, NH) ≤ 5. All phenolic acids examined herein meet the Lipinski's rule as demonstrated in Table 7. Thus, CinA, p-CoA and FA display favorable drug-like properties, which certainly encourage their further examination in the in vitro and in vivo studies.

Since phenolic compounds has been demonstrated as efficient cytostatic agents against various malignancies, it seems essential to evaluate how these compounds interact with lipid membranes, to get fuller insight into the transportation and anticancer mechanisms of these natural compounds. The affinity of polyphenols to the lipid bilayer is reflected in several electrochemical parameters. The first determinant is the adsorption on the membrane surface mediated by interactions of hydrophilic parts with the polar head groups of the lipids at the water–lipid interface. Second is the absorption dependent on the partitioning of the hydrophobic parts into the nonpolar core of the membrane [61,62]. The mechanism of action of polyphenols is determined by the presence of different substituents in their backbone structure and the pH value of their microenvironment. If the pH of the external medium is low (acidic), phenolic acids are then able to diffuse through the membrane due to their unchanged form [63,64]. The ELS data presented in Figures 2–4 clearly show that cinnamic acid and its derivatives do not affect the surface charge density of liposomal membranes at acidic pH. Therefore, it can be assumed that at acidic pH, investigated polyphenols were able to solubilize into the membrane and to permeate it. At neutral and basic pH, the CinA was unable to considerably modify the surface charge of the model membrane due to the lack of a hydroxyl group. Whereas more hydrophilic p-CoA and FA remained anchored at the bilayer surface without perturbing the lipidic structure but clearly affecting σ values. These findings are in agreement with literature, where the influence of structural characteristics of cinnamic acid and its hydroxyl derivatives on their interaction with model membranes was reported [51,65,66]. Similarly, the penetration of many other compounds e.g., flavonoids [7] or non-steroidal anti-inflammatory drugs [67] in bilayers, depends on the pH of the media.

Given the pK_{COOH} values (Table 7), at neutral and basic pH, the carboxyl group of CinA, p-CoA and FA is most likely negatively charged, and the negative charge of their molecules is probably oriented towards the positive pole of zwitterionic DOPC headgroup. ELS measurements performed in the presence of two concentrations of phenolic acids corroborate this hypothesis given that the surface charge of the membrane was altered, becoming increasingly negative with the addition of the examined polyphenols. The same tendency was observed for other negatively charged compounds [67]. The incorporation of cinnamic acid resulted in the slightest changes of σ, which again suggests interactions of CinA with the interior of the membrane. It is obvious that the effects of different phenolic acids are correlated to their structural characteristics, thereby even the difference in one –OH group can be important, as well as the number of H-bonds they form. CinA can form three H-bonds, with one as a H-bond donor, and two as H-bond acceptors; its partition coefficient was the highest, and it was the most soluble of these phenolic acids in octanol (Table 7). In other words, CinA as a non-polar substance could enter the lipid bilayer easily. Our findings are in line with the literature, where the interaction of model membranes with phenolic acids presenting different structural changes in their molecular backbone is deciphered on the basis of the shift of lipid phase transition temperature [51,61,66,68]. These authors reported that for pH below pK_{COOH} values, the protonation of the carboxylic group allows the substance to penetrate the lipid bilayer. Likewise, Castelli et al. employed multilamellar or unilamellar liposomes created from synthetic L-R-dimyristoylphosphatidylcholine to check whether cinnamic acid can dissolve into lipid membranes and penetrate them by migration from the aqueous phase [51]. This process continues until there is a constant molar fraction on the membrane surface, and then progressively inside the other internal bilayers. At the end of this process, the thermotropic behavior is close to that obtained by direct mixing of the biophenol with the lipid component during the liposomal preparation. The same researchers investigating PC liposomes containing p-CoA at two different pHs (4 and 7) reported no shift of the calorimetric peaks toward lower values, suggesting that the –OH group influences the ability of this compound to penetrate the membrane [51]. They did not exclude a surface interaction with the lipid layers and stated that the presence of different substituents in the backbone structure of biophenols might influence their incorporation. On the other hand, Ota et al. compared the thermograms obtained for the unilamelar large vesicles of 1,2-dipalmitoyl-sn-glycero-3-phosphocholine in the presence of the p-CoA or FA, and noticed that the enthalpy value of the main transition of the phospholipids decreased by 1.46 ± 0.10 kcal/mol in the presence of p-CoA, and by 0.45 ± 0.17 kcal/mol with FA. This small but significant decrease in the enthalpy of transition implies that both acids intercalate into the acyl chain region of the bilayer [66].

Analyzing the influence of the polar headgroup of lipids on the penetration capacity of cinnamic acid and its derivatives, we expected that phenolic acids will be less able to penetrate the negatively charged headgroup of PS lipids as compared with DOPC. Surprisingly, we have not noticed such an effect. We speculate that perhaps the acids concentrations used here were not sufficient to cause significant alterations of the membrane surface charge, or it may also be the result of certain limitations of the ELS technique per se. Likewise, Fadel et al. [64] reported that rosmarinic acid, another compound belonging to phenolic acids, evoked a weaker effect in PS than PC membranes.

Furthermore, we analyzed the influence of pH on the size of the liposomes. Our findings showed that for neutral liposomes (DOPC), their diameter increased with increasing pH of the solution. It may suggest that in acidic pH, existing electrostatic repulsive forces dominate over hydrogen bonding affinities between neighboring lipid molecules [69]. This might be a tentative explanation of why we observed smaller sizes of DOPC liposomes. Contrarily, for the negatively charged liposomes (DOPC/PS 9:1, DOPC/PS 8:2, PS), the diameter decreased with increasing pH. As such, at acidic pH, the extent of protonation was higher than this observed in the case of DOPC liposomes. Also, an opposite dependence, the dominance of hydrogen bonding over electrostatic repulsions, was observed. This might be the reason why the negatively charged liposomes showed larger sizes than neutral ones (DOPC). Simultaneously, in alkaline and neutral pH, reduction of the diameter of negatively charged liposomes was observed most probably due to the repulsive forces between phospholipid molecules [70].

The EIS data presented in Figures 6–8 and collected in Tables 4–6 indicate that the addition of cinnamic acid and its hydroxy derivatives to the DOPC, DOPC/PS 9:1 or DOPC/PS 8:2 membranes caused an increase in their resistance, and thus reduced conductivity. An increase in the R_m value implies that these phenolic acids exacerbated the ordering and decreased the dynamics of the phospholipid alkyl chains of spherical bilayers. This behavior can be explained in relation to the polarities of the examined molecules, indicating to what levels they can penetrate into the lipid bilayer. CinA, the least polar acid among the tested ones, had the greatest effect on the structure of the membrane lipids (i.e., stabilizing of the structure). *p*-CoA and FA are both more polar than CinA (see Figure 1), therefore, presenting a weaker effect on the membrane resistance. Based on our experimental approach, the following order of the stabilization effect was established: cinnamic acid > *p*-coumaric acid > ferulic acid. Our results are in line with previously reported data where the influence of the phenolic acids on structural properties of a model lipid membrane was investigated by differential scanning calorimetry [51,66,68], fluorescence spectroscopy [66,68], and electron paramagnetic resonance spectroscopy [68].

As opposed to the resistance, alterations in the capacitance seem to be less obvious in Nyquist plots, therefore, no remarkable differences between subsequent EIS measurements can be observed in Figures 6–8. It may be stated that the deposition of phenolic acids on the membrane was not accompanied by the appearance of any additional time constant, which was also reported not to happen in the case of protein adsorption [71]. Consequently, the C_m values fitted from the equivalent circuit model were in alignment with the indications from the Nyquist (Tables 4–6), revealing increasing capacitance with increasing phenolic acids concentrations.

Together, the EIS data indicate that in comparison to untreated plain phospholipid bilayer, CinA-stimulated membranes showed a significant and systematic increase in bilayer resistance, which is dependent on increasing the concentration of phenolic acid. These changes were accompanied by moderate elevation of the bilayer capacitance, which can be attributed to a decrease in membrane thickness. This in turn, may indicate the ordering and stabilizing effect of CinA on the phospholipid alkyl chains of bilayers.

According to the EIS response shown in Figures 6–8 and data collected in Tables 4–6, it is evident that *p*-CoA and FA interact with the bilayer in a different way than CinA. Both of these compounds elicit significant changes in the lipid membrane capacitance, which indicate their adsorption at the bilayer interface. In contrast to CinA, changes in membrane resistance caused by increasing concentrations of *p*-CoA or FA were much smaller. These observations were consistent with the ELS data, suggesting that the CinA associates strongly and penetrates deeply into the lipid membrane. Conversely, *p*-CoA and FA may locate near the phospholipid headgroup, most probably via electrostatic interaction, without perturbing the lipidic structure.

Finally, if we compare the influence of the polar headgroup of lipids on the R_m and C_m values of membranes modified with CinA, p-CoA or FA, it seems that phenolic acids are less able to interact with the negatively charged headgroup of PS lipids as compared with DOPC. This result is not surprising, because the data were registered at pH equals 6.59 (pH of 155 mM/L NaCl electrolyte solution), in which all tested compounds bear a negative charge (carboxylate group) that may cause charge repulsion between the two carboxylate functions of the serine in PS. Notably, we failed to identify the differences in the interaction between phenolic acids and DOPC or PS bilayers using zeta potential analysis. This indicates that impedance measurement is a useful and effective technique worth utilizing in physicochemical studies.

5. Conclusions

Currently, in the era of lipidomics, the attention of scientists from many fields has been shifted towards looking at biological membranes from a different perspective. Membranes have been demonstrated to determine certain physiological functions of cells and play important roles in several pathologies such as cancer. In this respect, a great deal of attention is now directed into the understanding of the interactions between anticancer drugs and cellular membranes. Thus, evaluation of drug-membrane dependencies can serve as a useful tool in predicting membrane permeability, bioactivity and cytotoxicity of potential antineoplastic agents in modern oncopharmacology. Regarding the importance of such interactions, the role of biomimetic model membranes and biophysical/electrochemical techniques becomes increasingly significant in chemical and pharmacological studies. As such, based on previous in vitro research conducted on glioblastoma cancer cells, we used electrophoretic light scattering and impedance spectroscopy to study the effects of CinA, *p*-CoA and FA on electrical properties of bilayers formed from DOPC, PS or DOPC-PS mixture. We demonstrated that after treatment with phenolic acids, the negative charge of membranes increased in alkaline pH solutions, but not in acidic ones. On the other hand, the data from impedance measurements showed elevated values of either the electrical capacitance and the electrical resistance in treated cells. We therefore concluded that at acidic pH, all tested compounds were able to solubilize into the membrane and permeate it. However, at neutral and alkaline pH, the CinA could be partially inserted into the bilayers, whereas *p*-CoA and FA could be anchored at the bilayer surface. Since intracellular penetration seems to be a key determinant of drug functioning, our results imply that electrochemical methods might be employed for predicting pharmacological activity and bioavailability of phenolic acids in the future.

Author Contributions: Conceptualization, M.N.; Methodology, J.K. and M.N.; Validation, J.K., M.G. and M.K.; Formal Analysis, J.K., M.Z., M.G. and M.N.; Investigation, M.Z.; Resources, M.N.; Data Curation, J.K. and M.G.; Writing—Original Draft Preparation, M.N.; Writing—Review and Editing, M.Z., J.K., M.K. and M.N.; Visualization, M.N., M.Z. and J.K.; Supervision, M.N.; Project Administration, M.N.; Funding Acquisition, M.N., M.G. All authors have read and agreed to the published version of the manuscript.

Funding: This study was funded by the National Science Centre (Poland) on the basis on the decision number 2018/02/X/ST4/02153 (Miniatura 2 call). This research was supported by the Slovak Research and Development Agency under the contract No. APVV-17-0149 and PP-COVID-20-0019. The potentiostat/galvanostat was funded by the European Funds for Regional Development and the National Funds of Ministry of Science and Higher Education as part of the Operational Program Development of Eastern Poland 2007–2013, project: POPW.01.03.00-20-044/11.

Conflicts of Interest: The authors declare no conflict of interest.

Abbreviations

ELS	Electrochemical light scattering
EIS	Electrochemical impedance spectroscopy
CinA	Cinnamic acid
p-CoA	*p*-Coumaric acid
FA	Ferulic acid
DOPC	1,2-Dioleoyl-sn-glycero-3-phosphocholine
PC	Phosphatidylcholine
PS	Phosphatidylserine
C_m	Membrane capacitance
R_m	Membrane resistance
σ	Surface charge density
$\log P$	n-octanol/water partition coefficient
clogP	Calculated n-octanol/water partition coefficient
Ro5	Rule of five
M_{WT}	Molecular weight
pK_a	Dissociation constant
pK_{COOH}	Dissociation constant of the carboxyl groups
pK_{OH}	Dissociation constant of the hydroxyl groups
HBA	Hydrogen bond acceptor groups
HBD	Hydrogen bond donor groups

References

1. Sokolová, R.; Degano, I.; Ramešová, Š.; Bulíčková, J.; Hromadová, M.; Gál, M.; Fiedler, J.; Valášek, M. The oxidation mechanism of the antioxidant quercetin in nonaqueous media. *Electrochim. Acta* **2011**, *56*, 7421–7427. [CrossRef]
2. Rice-Evans, C.A.; Miller, N.J.; Paganga, G. Structure-antioxidant activity relationships of flavonoids and phenolic acids. *Free. Radic. Biol. Med.* **1996**, *20*, 933–956. [CrossRef]

3. Apostolou, A.; Stagos, D.; Galitsiou, E.; Spyrou, A.; Haroutounian, S.; Portesis, N.; Trizoglou, I.; Hayes, A.W.; Tsatsakis, A.M.; Kouretas, D. Assessment of polyphenolic content, antioxidant activity, protection against ROS-induced DNA damage and anticancer activity of Vitis vinifera stem extracts. *Food Chem. Toxicol.* **2013**, *61*, 60–68. [CrossRef] [PubMed]

4. Lall, R.K.; Syed, D.N.; Adhami, V.M.; Khan, M.I.; Mukhtar, H. Dietary Polyphenols in Prevention and Treatment of Prostate Cancer. *Int. J. Mol. Sci.* **2015**, *16*, 3350–3376. [CrossRef] [PubMed]

5. Weng, C.-J.; Yen, G.-C. Chemopreventive effects of dietary phytochemicals against cancer invasion and metastasis: Phenolic acids, monophenol, polyphenol, and their derivatives. *Cancer Treat. Rev.* **2012**, *38*, 76–87. [CrossRef] [PubMed]

6. Zheng, X.; Cheng, Y.; Chen, Y.; Yue, Y.; Li, Y.; Xia, S.; Li, Y.; Deng, H.; Zhang, J.; Cao, Y.J. Ferulic Acid Improves Depressive-Like Behavior in Prenatally-Stressed Offspring Rats via Anti-Inflammatory Activity and HPA Axis. *Int. J. Mol. Sci.* **2019**, *20*, 493. [CrossRef] [PubMed]

7. Kruszewski, M.; Kusaczuk, M.; Kotyńska, J.; Gál, M.; Krętowski, R.; Cechowska-Pasko, M.; Naumowicz, M. The effect of quercetin on the electrical properties of model lipid membranes and human glioblastoma cells. *Bioelectrochemistry* **2018**, *124*, 133–141. [CrossRef]

8. Cárdenas, M.; Marder, M.; Blank, V.C.; Roguin, L.P. Antitumor activity of some natural flavonoids and synthetic derivatives on various human and murine cancer cell lines. *Bioorg. Med. Chem.* **2006**, *14*, 2966–2971. [CrossRef]

9. Li, H.; Liu, J.; Pei, T.; Bai, Z.; Han, R.; Liang, Z. Overexpression of SmANS Enhances Anthocyanin Accumulation and Alters Phenolic Acids Content in Salvia miltiorrhiza and Salvia miltiorrhiza Bge f. alba Plantlets. *Int. J. Mol. Sci.* **2019**, *20*, 2225. [CrossRef]

10. Kruszewski, M.A.; Kotyńska, J.; Kusaczuk, M.; Gál, M.; Naumowicz, M. The Modulating Effect of p-Coumaric Acid on The Surface Charge Density of Human Glioblastoma Cell Membranes. *Int. J. Mol. Sci.* **2019**, *20*, 5286. [CrossRef]

11. Zhang, X.; Lin, D.; Jiang, R.; Li, H.; Wan, J.; Li, H. Ferulic acid exerts antitumor activity and inhibits metastasis in breast cancer cells by regulating epithelial to mesenchymal transition. *Oncol. Rep.* **2016**, *36*, 271–278. [CrossRef] [PubMed]

12. Li, W.; Li, N.; Tang, Y.-P.; Li, B.; Liu, L.; Zhang, X.; Fu, H.; Duan, J.-A. Biological activity evaluation and structure–activity relationships analysis of ferulic acid and caffeic acid derivatives for anticancer. *Bioorg. Med. Chem. Lett.* **2012**, *22*, 6085–6088. [CrossRef] [PubMed]

13. Mori, H.; Kawabata, K.; Matsunaga, K.; Ushida, J.; Fujii, K.; Hara, A.; Tanaka, T.; Murai, H. Chemopreventive effects of coffee bean and rice constituents on colorectal carcinogenesis. *BioFactors* **2000**, *12*, 101–105. [CrossRef] [PubMed]

14. Ls, R.; Nja, S.; Ncp, S.; Mc, M.; Aj, T. Anticancer Properties of Phenolic Acids in Colon Cancer—A Review. *J. Nutr. Food Sci.* **2016**, *6*, 1000468. [CrossRef]

15. Serafim, T.L.; Carvalho, F.S.; Marques, M.P.M.; Calheiros, R.; Silva, T.; Garrido, J.M.; Milhazes, N.; Borges, F.; Roleira, F.M.; Silva, E.T.; et al. Lipophilic Caffeic and Ferulic Acid Derivatives Presenting Cytotoxicity against Human Breast Cancer Cells. *Chem. Res. Toxicol.* **2011**, *24*, 763–774. [CrossRef] [PubMed]

16. Dabrowska, M.; Starek, M.; Skucinski, J. Lipophilicity study of some non-steroidal anti-inflammatory agents and cephalosporin antibiotics: A review. *Talanta* **2011**, *86*, 35–51. [CrossRef] [PubMed]

17. Cumming, H.; Rücker, C. Octanol–Water Partition Coefficient Measurement by a Simple 1H NMR Method. *ACS Omega* **2017**, *2*, 6244–6249. [CrossRef]

18. Rocher, F.; Roblin, G.; Chollet, J.-F. Modifications of the chemical structure of phenolics differentially affect physiological activities in pulvinar cells of Mimosa pudica L. II. Influence of various molecular properties in relation to membrane transport. *Environ. Sci. Pollut. Res.* **2016**, *24*, 6910–6922. [CrossRef]

19. Lipinski, C.A.; Lombardo, F.; Dominy, B.W.; Feeney, P.J. Experimental and computational approaches to estimate solubility and permeability in drug discovery and development settings. *Adv. Drug Deliv. Rev.* **2012**, *64*, 4–17. [CrossRef]

20. Narayanaswamy, R.; Gnanamani, A.; Shanmugasamy, S.; Kumar, G.R.; Mandal, A.B. Bioinformatics in crosslinking chemistry of collagen with selective cross linkers. *BMC Res. Notes* **2011**, *4*, 399. [CrossRef]

21. Alam, A.; Tamkeen, N.; Imam, N.; Farooqui, A.; Ahmed, M.M.; Tazyeen, S.; Ali, S.; Malik, Z.; Ishrat, R. Pharmacokinetic and Molecular Docking Studies of Plant-Derived Natural Compounds to Exploring Potential Anti-Alzheimer Activity. In *Silico Approach for Sustainable Agriculture*; Springer: Singapore, 2018; pp. 217–238.

22. Van Balen, G.P.; Martinet, C.A.M.; Caron, G.; Reist, M.; Carrupt, P.-A.; Fruttero, R.; Gasco, A.; Testa, B. Liposome/water lipophilicity: Methods, information content, and pharmaceutical applications. *Med. Res. Rev.* **2004**, *24*, 299–324. [CrossRef] [PubMed]

23. New, R.R.C. *Liposomes. A Practical Approach*; IRL Press: Oxford, UK, 1990.

24. Liu, X.; Testa, B.; Fahr, A. Lipophilicity and Its Relationship with Passive Drug Permeation. *Pharm. Res.* **2010**, *28*, 962–977. [CrossRef] [PubMed]

25. McLaughlin, S. The Electrostatic Properties of Membranes. *Annu. Rev. Biophys. Biophys. Chem.* **1989**, *18*, 113–136. [CrossRef] [PubMed]

26. Smith, M.C.; Crist, R.M.; Clogston, J.D.; McNeil, S.E. Zeta potential: A case study of cationic, anionic, and neutral liposomes. *Anal. Bioanal. Chem.* **2017**, *409*, 5779–5787. [CrossRef]

27. Naumowicz, M.; Figaszewski, Z.A. Chronopotentiometric Technique as a Method for Electrical Characterization of Bilayer Lipid Membranes. *J. Membr. Biol.* **2011**, *240*, 47–53. [CrossRef]

28. Taylor, R.; Shults, J. (Eds.) *Handbook of Chemical and Biological Sensors*, 1st ed.; CRC Press: Boca Raton, FL, USA, 1996.

29. Khan, M.S.; Dosoky, N.S.; Berdiev, B.K.; Williams, J.D. Electrochemical impedance spectroscopy for black lipid membranes fused with channel protein supported on solid-state nanopore. *Eur. Biophys. J.* **2016**, *45*, 843–852. [CrossRef]

30. Velikonja, A.; Kramar, P.; Miklavčič, D.; Lebar, A.M. Specific electrical capacitance and voltage breakdown as a function of temperature for different planar lipid bilayers. *Bioelectrochemistry* **2016**, *112*, 132–137. [CrossRef]

31. Naumowicz, M.; Figaszewski, Z.A. Impedance Spectroscopic Investigation of the Bilayer Lipid Membranes Formed from the Phosphatidylserine–Ceramide Mixture. *J. Membr. Biol.* **2009**, *227*, 67–75. [CrossRef]

32. Brown, H. (Ed.) *Chemistry of the Cell Interface. Part A*, 1st ed.; Academic Press: New York, NY, USA, 1971; ISBN 9780323154581.

33. Coster, H.G. Dielectric and Electrical Properties of Lipid Bilayers in Relation to their Structure. *Membr. Sci. Technol.* **2003**, *7*, 75–108. [CrossRef]

34. Kotyńska, J.; Figaszewski, Z.A. Microelectrophoretic investigation of the interactions between liposomal membranes formed from a phosphatidylcholine-phosphatidylglycerol mixture and monovalent ions. *Eur. Phys. J. E* **2014**, *37*, 1–6. [CrossRef]

35. Naumowicz, M.; Figaszewski, Z. Pore Formation in Lipid Bilayer Membranes made of Phosphatidylcholine and Cholesterol Followed by Means of Constant Current. *Cell Biophys.* **2012**, *66*, 109–119. [CrossRef] [PubMed]

36. Naumowicz, M.; Petelska, A.D.; Figaszewski, Z. Impedance Analysis of Complex Formation Equilibria in Phosphatidylcholine Bilayers Containing Decanoic Acid or Decylamine. *Cell Biophys.* **2011**, *61*, 145–155. [CrossRef] [PubMed]

37. Naumowicz, M.; Kruszewski, M.A.; Gál, M. Electrical properties of phosphatidylcholine bilayers containing canthaxanthin or β-carotene, investigated by electrochemical impedance spectroscopy. *J. Electroanal. Chem.* **2017**, *799*, 563–569. [CrossRef]

38. Svennerholm, L. Distribution and fatty acid composition of phosphoglycerides in normal human brain. *J. Lipid Res.* **1968**, *9*, 570–579. [PubMed]

39. Bangham, A.; Standish, M.; Watkins, J. Diffusion of univalent ions across the lamellae of swollen phospholipids. *J. Mol. Biol.* **1965**, *13*, 238–252. [CrossRef]

40. Naumowicz, M.; Kusaczuk, M.; Zając, M.; Gál, M.; Kotyńska, J. Monitoring of the Surface Charge Density Changes of Human Glioblastoma Cell Membranes upon Cinnamic and Ferulic Acids Treatment. *Int. J. Mol. Sci.* **2020**, *21*, 6972. [CrossRef]

41. Naumowicz, M.; Kusaczuk, M.; Kruszewski, M.A.; Gál, M.; Krętowski, R.; Cechowska-Pasko, M.; Kotyńska, J. The modulating effect of lipid bilayer/p-coumaric acid interactions on electrical properties of model lipid membranes and human glioblastoma cells. *Bioorg. Chem.* **2019**, *92*, 103242. [CrossRef]

42. Makroaufmassprogram Program. Available online: http://ruedig.de/tmp/messprogramm.htm (accessed on 22 March 2020).

43. Alexander, A.E.; Johnson, P. *Colloid Science*; Clarendon Press: Oxford, UK, 1949.

44. Barrow, G.M. *Physical Chemistry*; McGraw-Hill: New York, NY, USA, 1996.

45. Naumowicz, M.; Figaszewski, Z. Impedance Analysis of Phosphatidylcholine/α-Tocopherol System in Bilayer Lipid Membranes. *J. Membr. Biol.* **2005**, *205*, 29–36. [CrossRef]

46. Laszuk, P.; Urbaniak, W.; Petelska, A.D. The Equilibria in Lipid–Lipoic Acid Systems: Monolayers, Microelectrophoretic and Interfacial Tension Studies. *Molecules* **2020**, *25*, 3678. [CrossRef]

47. Zhou, Y.; Raphael, R.M. Solution pH Alters Mechanical and Electrical Properties of Phosphatidylcholine Membranes: Relation between Interfacial Electrostatics, Intramembrane Potential, and Bending Elasticity. *Biophys. J.* **2007**, *92*, 2451–2462. [CrossRef]

48. Ramadurai, S.; Sarangi, N.K.; Maher, S.; MacConnell, N.; Bond, A.M.; McDaid, D.; Flynn, D.; Keyes, T.E. Microcavity-Supported Lipid Bilayers; Evaluation of Drug–Lipid Membrane Interactions by Electrochemical Impedance and Fluorescence Correlation Spectroscopy. *Langmuir* **2019**, *35*, 8095–8109. [CrossRef] [PubMed]

49. Naumowicz, M.; Figaszewski, Z.A. Chronoamperometric Technique as a Useful Method for Electrical Characterization of Bilayer Lipid Membranes. *J. Electrochem. Soc.* **2013**, *161*, H11–H16. [CrossRef]

50. Dávalos, J.Z.; Herrero, R.; Chana, A.; Guerrero, A.; Jiménez, P.; Santiuste, J.M. Energetics and Structural Properties, in the Gas Phase, of trans-Hydroxycinnamic Acids. *J. Phys. Chem. A* **2012**, *116*, 2261–2267. [CrossRef] [PubMed]

51. Castelli, F.; Uccella, N.A.; Trombetta, D.; Saija, A. Differences between coumaric and cinnamic acids in membrane permeation as evidenced by time-dependent calorimetry. *J. Agric. Food Chem.* **1999**, *47*, 991–995. [CrossRef]

52. Silva, F.A.M.; Borges, F.; Guimarães, C.; Lima, J.L.F.C.; Matos, C.; Reis, S. Phenolic Acids and Derivatives: Studies on the Relationship among Structure, Radical Scavenging Activity, and Physicochemical Parameters. *J. Agric. Food Chem.* **2000**, *48*, 2122–2126. [CrossRef]

53. Kealey, D. *Experiments in Modern Analytical Chemistry*; Springer Science and Business Media: New York, NY, USA, 1986.

54. Galanakis, C.; Goulas, V.; Tsakona, S.; Manganaris, G.; Gekas, V. A Knowledge Base for The Recovery of Natural Phenols with Different Solvents. *Int. J. Food Prop.* **2013**, *16*, 382–396. [CrossRef]

55. Jablonsky, M.; Haz, A.; Burcova, Z.; Kreps, F.; Jablonsky, J. Pharmacokinetic properties of biomass-extracted substances isolated by green solvents. *BioResources* **2019**, *14*, 6294–6303. [CrossRef]

56. Naseem, S.; Laurent, A.D.; Carroll, E.C.; Vengris, M.; Kumauchi, M.; Hoff, W.D.; Krylov, A.I.; Larsen, D.S. Photo-isomerization upshifts the pKa of the Photoactive Yellow Protein chromophore to contribute to photocycle propagation. *J. Photochem. Photobiol. A Chem.* **2013**, *270*, 43–52. [CrossRef]

57. Friedman, M.; Jürgens, H.S. Effect of pH on the Stability of Plant Phenolic Compounds. *J. Agric. Food Chem.* **2000**, *48*, 2101–2110. [CrossRef]

58. Alves, A.C.; Ribeiro, D.; Horta, M.; Lima, J.L.F.C.; Nunes, C.; Reis, S. A biophysical approach to daunorubicin interaction with model membranes: Relevance for the drug's biological activity. *J. R. Soc. Interface* **2017**, *14*, 20170408. [CrossRef]

59. Ramalho, M.J.; Andrade, S.; Coelho, M.; Álvaro, N.; Loureiro, J.A.; Pereira, M.C. Biophysical interaction of temozolomide and its active metabolite with biomembrane models: The relevance of drug-membrane interaction for Glioblastoma Multiforme therapy. *Eur. J. Pharm. Biopharm.* **2019**, *136*, 156–163. [CrossRef] [PubMed]

60. Cheynier, V.; Comte, G.; Davies, K.M.; Lattanzio, V.; Martens, S. Plant phenolics: Recent advances on their biosynthesis, genetics, and ecophysiology. *Plant Physiol. Biochem.* **2013**, *72*, 1–20. [CrossRef] [PubMed]

61. Ulrih, N.P. Effects of Caffeic, Ferulic, and p-Coumaric Acids on Lipid Membranes. In *Coffee in Health and Disease Prevention*; Preedy, V.R., Ed.; Academic Press: San Diego, CA, USA, 2015; pp. 813–821, ISBN 9780124167162.

62. Sirk, T.W.; Brown, E.F.; Sum, A.K.; Friedman, M. Molecular Dynamics Study on the Biophysical Interactions of Seven Green Tea Catechins with Lipid Bilayers of Cell Membranes. *J. Agric. Food Chem.* **2008**, *56*, 7750–7758. [CrossRef] [PubMed]

63. Damasceno, S.S.; Dantas, B.B.; Ribeiro-Filho, J.; Araújo, D.A.M.; Da Costa, J.G.M. Chemical Properties of Caffeic and Ferulic Acids in Biological System: Implications in Cancer Therapy. A Review. *Curr. Pharm. Des.* **2017**, *23*, 3015–3023. [CrossRef]

64. Fadel, O.; El Kirat, K.; Morandat, S. The natural antioxidant rosmarinic acid spontaneously penetrates membranes to inhibit lipid peroxidation in situ. *Biochim. Biophys. Acta (BBA) Biomembr.* **2011**, *1808*, 2973–2980. [CrossRef]

65. Phan, H.T.; Yoda, T.; Chahal, B.; Morita, M.; Takagi, M.; Vestergaard, M. Structure-dependent interactions of polyphenols with a biomimetic membrane system. *Biochim. Biophys. Acta (BBA) Biomembr.* **2014**, *1838*, 2670–2677. [CrossRef]

66. Ota, A.; Abramovič, H.; Abram, V.; Ulrih, N.P. Interactions of p-coumaric, caffeic and ferulic acids and their styrenes with model lipid membranes. *Food Chem.* **2011**, *125*, 1256–1261. [CrossRef]
67. Lúcio, M.; Ferreira, H.; Lima, J.L.F.C.; Reis, S. Use of liposomes as membrane models to evaluate the contribution of drug–membrane interactions to antioxidant properties of etodolac. *Redox Rep.* **2008**, *13*, 225–236. [CrossRef]
68. Ulrih, N.P.; Ota, A.; Abram, V. Impact of selected polyphenolics on the structural properties of model lipid membranes—A review. *Int. J. Food Stud.* **2017**, *6*, 158–177. [CrossRef]
69. Sułkowski, W.; Pentak, D.; Nowak, K.; Sułkowska, A. The influence of temperature, cholesterol content and pH on liposome stability. *J. Mol. Struct.* **2005**, *744*, 737–747. [CrossRef]
70. Roy, B.; Guha, P.; Bhattarai, R.; Nahak, P.; Karmakar, G.; Chettri, P.; Panda, A.K. Influence of Lipid Composition, pH, and Temperature on Physicochemical Properties of Liposomes with Curcumin as Model Drug. *J. Oleo Sci.* **2016**, *65*, 399–411. [CrossRef] [PubMed]
71. Dutoit, M.; Ngaboyamahina, E.; Wiesner, M. Pairing electrochemical impedance spectroscopy with conducting membranes for the in situ characterization of membrane fouling. *J. Membr. Sci.* **2021**, *618*, 118680. [CrossRef]

Publisher's Note: MDPI stays neutral with regard to jurisdictional claims in published maps and institutional affiliations.

 *membranes*

Article

Electrochemical Properties of Lipid Membranes Self-Assembled from Bicelles

Damian Dziubak [1,*], Kamil Strzelak [2] and Slawomir Sek [1]

1   Faculty of Chemistry, Biological and Chemical Research Centre, University of Warsaw, Żwirki i Wigury 101, 02-089 Warsaw, Poland; slasek@chem.uw.edu.pl
2   Faculty of Chemistry, University of Warsaw, Pasteura 1, 02-093 Warsaw, Poland; kamil.strzelak@chem.uw.edu.pl
*   Correspondence: ddziubak@chem.uw.edu.pl

Abstract: Supported lipid membranes are widely used platforms which serve as simplified models of cell membranes. Among numerous methods used for preparation of planar lipid films, self-assembly of bicelles appears to be promising strategy. Therefore, in this paper we have examined the mechanism of formation and the electrochemical properties of lipid films deposited onto thioglucose-modified gold electrodes from bicellar mixtures. It was found that adsorption of the bicelles occurs by replacement of interfacial water and it leads to formation of a double bilayer structure on the electrode surface. The resulting lipid assembly contains numerous defects and pinholes which affect the permeability of the membrane for ions and water. Significant improvement in morphology and electrochemical characteristics is achieved upon freeze–thaw treatment of the deposited membrane. The lipid assembly is rearranged to single bilayer configuration with locally occurring patches of the second bilayer, and the number of pinholes is substantially decreased. Electrochemical characterization of the lipid membrane after freeze–thaw treatment demonstrated that its permeability for ions and water is significantly reduced, which was manifested by the relatively high value of the membrane resistance.

Keywords: electrochemistry; gold electrode; supported lipid membranes; bicelles; self-assembly

Citation: Dziubak, D.; Strzelak, K.; Sek, S. Electrochemical Properties of Lipid Membranes Self-Assembled from Bicelles. *Membranes* **2021**, *11*, 11. https://dx.doi.org/10.3390/membranes11010011

Received: 1 December 2020
Accepted: 21 December 2020
Published: 23 December 2020

Publisher's Note: MDPI stays neutral with regard to jurisdictional claims in published maps and institutional affiliations.

1. Introduction

Lipid bilayers supported on solid substrates are considered as an important model to mimic the natural cell membranes in fundamental studies [1–4]. These systems were also proven to be suitable for the construction of biosensors and bioanalytical platforms for the examination of membrane proteins [5–7]. The immobilization of the membrane at the supporting substrate offers a unique opportunity to probe the properties of such an assembly with numerous surface-sensitive techniques. These include scanning probe microscopy, infrared reflective absorption spectroscopy, quartz crystal microbalance, and for conductive supports, electrochemical methods can be used as well [8–11]. Most popular approaches for supported lipid membrane formation involve vesicles spreading or Langmuir-Blodgett and Langmuir-Schafer techniques [12–15]. It was demonstrated in numerous research papers that both can produce well-defined planar bilayers with good electrical insulating properties manifested by low differential capacitance and high membrane resistance [7,16]. This issue is of crucial importance for the biosensors, which are sensitive to the structural or functional changes of lipid assemblies triggered by membrane proteins, or the constructs, which may act as affinity sensors detecting interactions of biological material with lipid membrane [17,18]. In particular, the modulation of the ion permeability of lipid membranes may be utilized in biosensors or the studies of pore-forming toxins since the dielectric damage can be transduced into the physical signal utilizing electrochemical methods [16].

An alternative approach was proposed involving the formation of supported lipid membranes from bicellar mixtures. Bicelles are composed of long-chain and short-chain phospholipids and tend to form disk-like aggregates [19,20]. Such lipidic assemblies are broadly utilized in structural biology studies due to their ability to host membrane proteins while retaining protein structure and function [21]. The planar bilayer formation on silicon chips from a bicellar mixture of 1,2-dipalmitoyl-*sn*-glycero-3-phosphocholine and 1,2-diheptanoyl-*sn*-glycero-3-phosphocholine lipids was first reported by Zeineldin and coworkers [22]. The optimal routes to fabricate supported lipid bilayers from bicelles were later demonstrated by Cho's group, who also revealed mechanistic details of the lipid membrane formation on hydrophilic substrates such as silicon dioxide, titanium oxide, and aluminum oxide [23,24]. These authors have shown that the adsorption behavior of bicelles can vary depending on the nature of the supporting surface, and the electrostatic attraction between the surface and adsorbing bicelles is necessary for the successful formation of the supported lipid bilayer (SLB). A similar conclusion can be drawn based on the results reported by Yamada and coworkers, who utilized atomic force microscopy and force spectroscopy to probe the properties of lipid films assembled from bicelles [25]. It was found that lamellae of phospholipid bilayers were aligned parallel to a surface in case of the negatively charged bare silicon substrate, while unoriented phospholipid bilayers were formed on Si substrate modified with terminal amine groups, where the excess of positive surface charge is expected.

In this work, we have described the mechanism of bicelles adsorption onto thioglucose-modified gold electrodes. The architecture of the resulting membrane can be considered as a floating lipid membrane, which is separated from the substrate by a monolayer of hydrophilic molecules of thioglucose [26,27]. In such a configuration, the polar heads of lipids located close to the electrode surface remain hydrated and the direct interaction with metal is eliminated. Immobilization of the lipid assembly on conductive support enabled electrochemical characterization of the resulting membrane and assessment of its permeability for ions. Additionally, the effect of freeze–thaw treatment on membrane electrical insulating properties and morphology was investigated.

2. Materials and Methods

Chemicals. 1,2-dimyristoyl-*sn*-glycero-3-phosphocholine (DMPC) and 3-([3-cholamidopropyl]dimethylammonio)-2-hydroxy-1-propanesulfonate (CHAPSO) were purchased from Avanti Polar Lipids Inc., Alabaster, AL, USA. Sodium fluoride, 1-thio-β-D-glucose, hydrofluoric acid, and sodium tetrachloroaurate were purchased from Sigma-Aldrich Sp. z. o. o., Poznan, Poland. All other reagents and solvents were obtained from Avantor Performance Materials Poland S.A., Gliwice, Poland. All chemicals were used as received. The water was purified through the Milli-Q system (resistivity 18.2 MΩ × cm). In all experiments, we have used an aqueous solution of 0.1 M NaF.

Bicelles preparation. Bicelles were prepared according to the protocol described by Ujwal and coworkers [28]. 260 mg of DMPC and 90 mg of CHAPSO were transferred to the vial and mixed with 1 mL of ultrapure Milli-Q water with a resistivity of 18.2 MΩ × cm. To disperse lipids and obtain bicelles, the suspension was warmed to 40 °C and then cooled to −20 °C. The steps with warming and cooling were repeated more than 25 times until the mixture at room temperature was homogeneous and in gel consistency. However, 10 cycles should be enough to observe the formation of the gel at room temperature and cloudiness upon cooling. For the experiments, bicelles were dissolved in 0.1 M NaF solution with a volume ratio of 1:1000.

Electrochemistry. Electrochemical experiments were performed using CHI 650B potentiostat (CH Instruments Inc., Austin, TX, USA) in a three-electrode cell with an Ag | AgCl | sat.KCl reference electrode, a Pt foil counter electrode, and an Au(111) working electrode. The measurements were carried out in a hanging meniscus configuration. Before each experiment, the Au(111) working electrode was cleaned in piranha solution (H$_2$SO$_4$: H$_2$O$_2$ 3:1 *v/v*. CAUTION: piranha reacts violently with organic compounds) for at least 12 h. Then it was thoroughly rinsed with ultrapure water and flame annealed. The monolayers of 1-thio-β-D-glucose (further referenced as thioglucose) were obtained by immersing the Au(111) electrode in an aqueous solution containing 0.1 mg/mL of the thioglucose for ~2 h. Next, the electrodes were rinsed with ultrapure water and immersed in bicellar suspension for at least 3 h to obtain lipid assembly. For the freeze–thaw treatment, the electrode was taken out from the suspension and covered with a thin film of the supporting electrolyte. Freeze–thaw treatment was carried out by cooling the electrode with deposited lipid film down to $-18\,^{\circ}\mathrm{C}$ for 2 h and then warming it up to room temperature. A single freeze–thaw cycle was performed for each experiment. AC voltammetry measurements were performed with a scan rate of 5 mV/s. The RMS amplitude of AC perturbation was 10 mV, and a frequency set at 20 Hz. The differential capacitance was derived based on AC voltammograms from the in-phase and out-of-phase components of the AC signal under the assumption that the electrode-electrolyte interface can be treated as a simple RC circuit. Electrochemical impedance spectroscopy (EIS) measurements were carried out within the frequency range of 10^{-1} to 10^4 Hz and the amplitude AC perturbation was 10 mV. All measurements were carried at $22 \pm 1\,^{\circ}\mathrm{C}$. The potentials reported in this work are referenced to the Ag | AgCl | sat.KCl electrode.

Topography Imaging. Atomic force microscopy (AFM) experiments were performed with Dimension Icon (Bruker Corporation, Billerica, MA, USA). The imaging of the samples was performed using ScanAsyst Fluid probes (Bruker, nominal spring constant 0.7 N/m, tip radius ~20 nm) in PeakForce Tapping mode. The cantilever was periodically modulated with default amplitude at the frequency of 2 kHz. The exact value of the spring constant and the deflection sensitivity for a given probe was carefully calibrated by thermal tuning before each experiment. All images were recorded in an aqueous solution of 0.1 M NaF at the temperature of $22 \pm 1\,^{\circ}\mathrm{C}$. In situ imaging was performed on an Au(111) single crystal (MaTecK, GmbH, Julich, Germany) pre-modified with thioglucose monolayer. The quality of the bare Au(111) surface was verified by AFM imaging before each experiment (see Figure S1 in Supplementary Materials). The images were processed using Nanoscope Analysis software version 1.40 (Bruker Corporation, Billerica, MA, USA) and involved flattening with a first-order polynomial function. The protocol of the thioglucose immobilization on gold was the same as for electrochemical measurements. The images were collected immediately after injection of bicellar suspension into the liquid cell of the atomic force microscope. Freeze–thaw treatment was performed by cooling the sample with deposited lipid film down to $-18\,^{\circ}\mathrm{C}$ and then warming it up to room temperature.

Surface Enhanced Infrared Absorption Spectroscopy. The spectra were recorded with Nicolet iS50 FTIR spectrometer (Thermo Fisher Scientific, Waltham, MA, USA) with MCT-A detector and custom-made single-reflection accessory. The incident angle was 60° and the spectral resolution was 4 cm^{-1}. The all-glass custom-made spectroelectrochemical cell was used in all experiments with platinum foil serving as a counter electrode and Ag | AgCl | sat.KCl as a reference electrode. The working electrode was a thin gold film deposited on a reflectance plane of a Si hemispherical prism. Deposition of gold was carried out by dropping an aqueous plating solution onto the hydrogen-terminated Si surface. The plating solution was obtained by mixing 100 μL of 0.03 M NaAuCl$_4 \cdot$2H$_2$O, 2 mL of 0.15 M Na$_2$SO$_3$ + 0.05 M Na$_2$S$_2$O$_3 \cdot$5H$_2$O + 0.05 M NH$_4$Cl, and 1 mL of 2% HF. After approximately 90 s of plating, the prism was rinsed with ultrapure water to finish the deposition. The deposited film was further modified by dropping an aqueous solution of thioglucose (0.1 mg/mL) onto the gold surface and after 2 h the prism was gently rinsed with ultrapure water. The spectra are displayed in absorbance units defined as A = $\log(I_0/I)$, where I_0 corresponds to the intensities of IR radiation observed for the reference spectra, while I corresponds to the intensity observed for the sample. The reference spectrum was collected for gold film modified with thioglucose. Data processing was performed using Omnic 9 software (Thermo Fisher Scientific, Waltham, MA, USA).

3. Results and Discussion

Self-assembly of bicelles on thioglucose-modified gold was monitored using surface-enhanced infrared absorption spectroscopy (SEIRAS). In this technique, the infrared absorption intensity can be significantly enhanced by 10–1000 times on coinage metal nanoparticles [29]. The electromagnetic field of the incident light induces an oscillating dipole in the metal nanoparticle by excitation of the localized plasmon. The induced dipole produces an electric field in the vicinity of the nanoparticle and excites the absorbed molecules. The electric field is normal to the metal surface, which means that molecular vibrations with the component of the transition dipole moment normal to the local surface can be excited and enhanced. Moreover, the local electric field decays within a short distance away from the surface. Since the dominant contribution comes from the species close to the metal surface, the adsorption and desorption processes can be monitored using this technique. Figure 1 illustrates the time evolution of the spectra recorded during DMPC-CHAPSO bicelles deposition. The spectra were collected at open circuit potential, and each represents the difference in absorbance between the spectrum at the specified time and the spectrum recorded as reference. The latter was collected for thioglucose-modified gold film deposited on a silicon prism before the addition of bicellar suspension. The most pronounced changes occur in the spectral region corresponding to the O–H stretching between 3000 and 3600 cm^{-1}. The absolute intensity of the negative band increases regularly within the time, demonstrating that the amount of water in the interfacial region decreases gradually. The minimum of the v(O–H) band is located at ~3225 cm^{-1} and it contains a visible shoulder at ~3350 cm^{-1}. The v(O–H) bands within this spectral region correspond to the water molecules in a network of hydrogen bonds [30–32]. Hence, the emergence of the negative v(O–H) band indicates that the adsorption of bicelles involves the replacement of hydrogen-bonded water from the interfacial region. This is confirmed by the presence of the negative δ(O–H) band at ~1630 cm^{-1}, which can also be ascribed to hydrogen-bonded water.

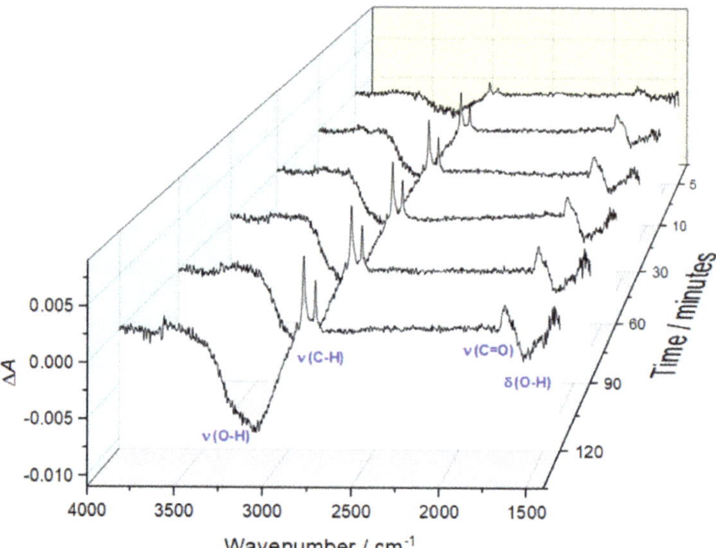

Figure 1. Time-evolution of surface-enhanced infrared absorption (SEIRA) spectra collected for thioglucose-modified gold during the deposition of bicelles. The reference spectrum was collected for gold film modified with monolayer of thioglucose.

In parallel with the loss of interfacial water, an increase in the intensity of the positive bands corresponding to v(C–H) and ester v(C=O) vibrations is observed. Both are related to the adsorption of lipid molecules on the electrode surface. The bands associated with v_{as}(C–H) and v_s(C–H) vibrations are located at 2917 cm^{-1} and 2849 cm^{-1}, respectively, which indicates that the lipids forming the assembly are in the gel state with the acyl chains adopting all-trans conformation [33]. This observation is understandable since the experiments were performed at 22 °C and the main gel (L_β) to liquid–crystalline (L_α) phase transition of DMPC occurs at ~23–24 °C. Importantly, the position of the ester ν(C=O) band at ~1725 cm^{-1} is indicative of hydrogen-bonded populations of ester groups that are fully hydrated [34]. These experimental observations demonstrate that bicelles are successfully deposited on thioglucose-modified gold, and the effect of the substrate on lipid conformation and hydration of the polar heads is rather small.

Further evaluation of bicelles adsorption was performed using in situ atomic force microscopy (AFM). This method has a unique capability to image the topography of the surface films in nanoscale and to provide information about their thickness. Importantly, the imaging can be performed under in situ conditions, therefore numerous surface-related processes can be monitored in real-time [11,35,36].

Figure 2 presents the images of a thioglucose-modified Au(111) electrode exposed to bicellar suspension. Within approximately 10 min upon injection of the bicelles, the electrode surface is covered with patches of lipidic material with the height varying within the range of ~10–15 nm. Since the expected thickness of the single bilayer formed by DMPC is ~4–5 nm, this indicates that bicelles tend to adsorb as stacks and form double or triple bilayers. Time-lapse imaging revealed that randomly shaped lipidic deposits further grow and merge into the stable planar film with numerous pinholes. The total thickness of the film measured from cross-sectional profiles taken along large defect sites was found to be 9.5 $\pm$ 0.4 nm, indicating that the lipid assembly is a double bilayer, and the defects span the entire thickness of the film. However, the depth of some fraction of the smaller pinholes was 4.8 $\pm$ 0.4 nm (see Figure 2C). In such a case, only the top bilayer is perforated, while the bottom film maintains continuity. Thus, the AFM imaging confirmed successful adsorption of bicelles on a thioglucose-modified gold surface. However, the resulting film does not adopt a single bilayer configuration, and it contains numerous defects spanning either the entire film or only the top bilayer.

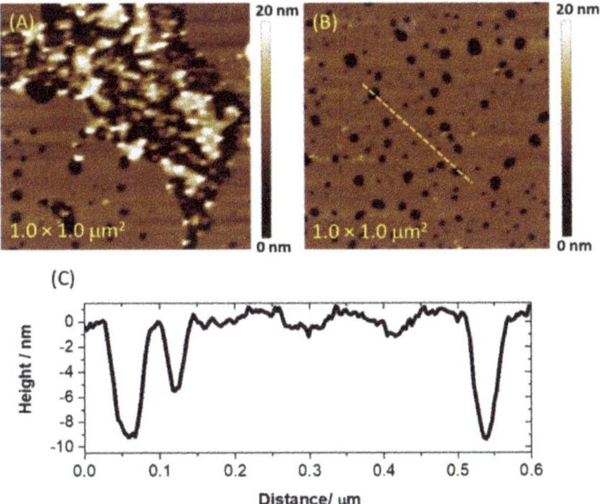

Figure 2. Atomic force microscopy (AFM) images collected for thioglucose-modified Au(111) after (**A**) 10 min and (**B**) 90 min of deposition of bicelles. Panel (**C**) presents the cross-sectional profile for the completed lipid assembly. The profile was taken along the orange dotted line displayed in panel (**B**). The images were collected in Peak Force Tapping mode.

To improve the properties of the lipid assembly, we have subjected the samples to freeze–thaw cycle. Such an approach assumes that osmotic stress during freezing and thawing can lead to the destabilization and mechanical rupture of lipid membranes, which in our case might induce significant membrane reorganization. Moreover, the previous AFM studies have demonstrated that interfacial water in phosphatidylcholine lipid bilayer is ordered at room temperature and the ions act as bridges between the lipid polar heads [37,38]. However, when the temperature is lowered below 4 °C, the order of the hydration shell of the bilayer might decrease due to the negative thermal expansion coefficient of water. This is manifested by the substantial decrease of Young's modulus of the bilayer, which could be related to increased efflux of bridging ions and the creation of a thicker and more disordered interface. As a result, the distance between polar heads is increased, which could make the assembly more prone to reorganization. The effect of the freeze–thaw cycle on the topography of the lipid film is shown in Figure 3.

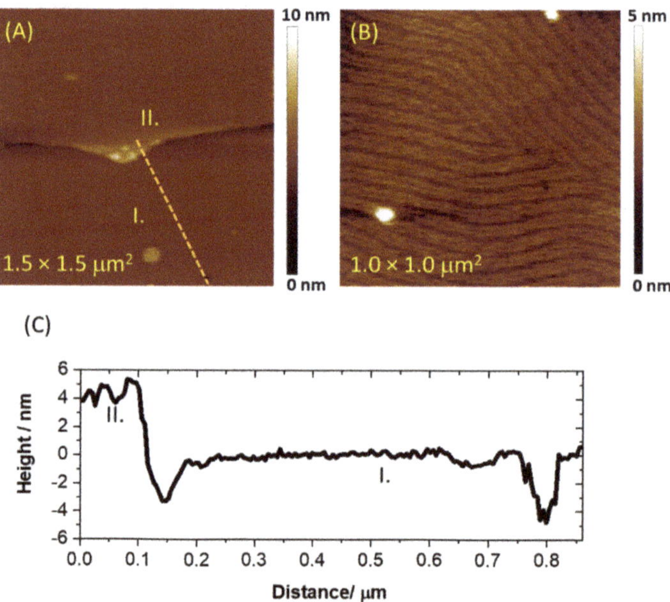

Figure 3. (**A**) An AFM image collected for lipid assembly deposited on thioglucose-modified Au(111) after one cycle of freeze–thaw treatment. (**B**) An AFM image of the double bilayer region with ripple phase structure. (**C**) The cross-sectional profile for lipid assembly after one freeze–thaw cycle. The profile was taken along the orange dotted line displayed in panel (**A**). The images were collected in Peak Force Tapping mode.

In this case, the number of the pinholes is significantly reduced, and the large area of the electrode is covered with continuous film represented by region I in Figure 3A. Its thickness was found to be 4.7 ± 0.3 nm, proving that a single bilayer is formed upon the freeze–thaw cycle. Nevertheless, the presence of the double bilayer is still observed locally, which is represented by region II in Figure 3A. Interestingly, the top bilayer in region II exhibits corrugation with well-defined periodicity typical for the ripple phase ($P_{\beta'}$) [39,40]. Figure 3B shows a more detailed image of this structure, and it can be observed that the corrugations form long ripples composed of twinned stripes. The spacing between the neighboring ripples is 54 ± 10 nm, while the distance between twinned lines is 15 ± 3 nm (see Figure S2 in Supplementary Materials). The average amplitude of the ripples was found to be 0.92 ± 0.18 nm. The ripple phase is known to exist in a temperature range between the pretransition and the main phase transition, and it is usually observed for multibilayer phospholipid assemblies supported on a solid surface [40]. Importantly, the repetitive AFM imaging of the samples stored at 22 °C under the electrolyte solution revealed that the morphology of the lipid film does not change significantly within one week. The only visible change was related to the gradual disappearance of the ripple phase.

Hence, the AFM data demonstrate clearly that the perforated double bilayer formed by bicelles deposition on thioglucose-modified gold can be effectively changed by the freeze–thaw procedure. The latter leads to the formation of a single bilayer coexisting with a double bilayer, and the number of pinholes is significantly reduced. Such a transformation should affect the electrochemical behavior of the lipid film. Therefore, we have used AC voltammetry and electrochemical impedance spectroscopy to further investigate the properties of the lipid membranes. Figure 4 illustrates the changes in a differential capacitance as a function of the potential recorded for thioglucose-modified gold electrodes with deposited lipid film.

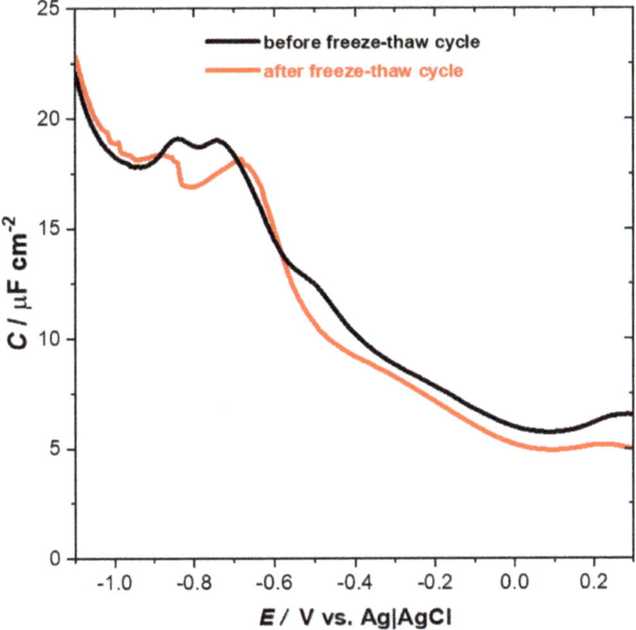

Figure 4. Potential dependence of differential capacitance of the thioglucose-modified Au(111) electrode with lipid film before (black) and after (red) one freeze–thaw cycle.

Potential-dependent changes in differential capacitance are quite similar before and after freeze–thaw treatment. The lipid film is most stable between +0.2 V and −0.1 V, and a well-pronounced minimum of the capacitance is observed at +0.1 V. Negative polarization of the electrode causes an increase of the capacitance, which is often explained by electroporation and/or lifting the membrane leading to accumulation of water and ions at the interfacial region [31,41]. At the potential of −0.6 V, the pseudocapacitive peaks are observed reflecting the desorption of the lipid film from the electrode, which is then followed by reductive desorption of thioglucose monolayer at −0.8 V. Although, the observed changes are qualitatively similar, it is evident that the properties of the lipid film that after freeze–thaw cycle are improved, as it is manifested by an overall decrease of the measured capacitance. Referring to AFM data, this effect can be ascribed to fewer pinholes and defect sites after freeze–thaw treatment. Further verification of the electrical properties of the lipid membrane assembled from bicelles was carried out using electrochemical impedance spectroscopy (EIS). The resulting Bode plots are shown in Figure 5. The spectra were collected at the potential of +0.1 V, where a minimum in differential capacitance is observed and the lipid membrane is most stable.

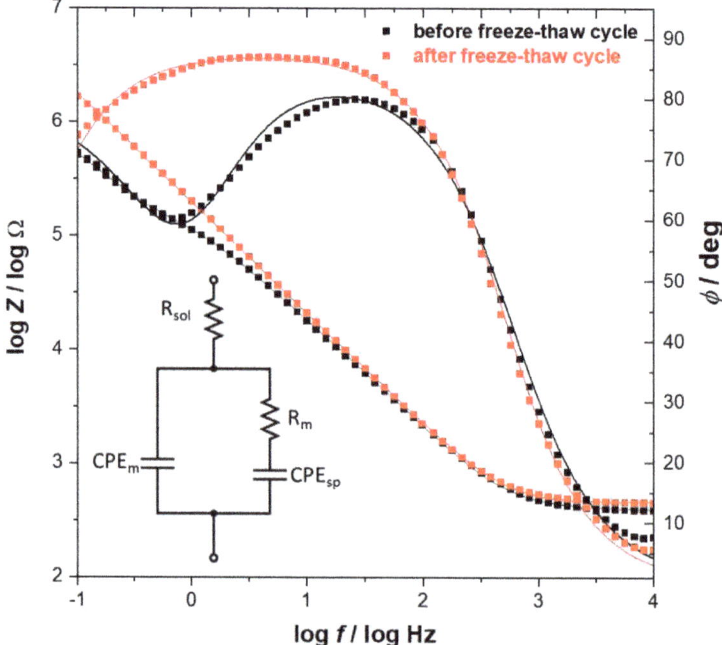

Figure 5. Bode plot obtained for the thioglucose-modified Au(111) electrode with lipid film before (black) and after (red) one freeze-thaw cycle. Solid lines represent the fit based on the equivalent circuit shown as an inset, where R_{sol} is the resistance of the electrolyte, R_m and CPE_m are the resistance and constant phase element of the lipid membrane, and CPE_{sp} represents the constant phase element of the spacer region between the gold surface and lipid film.

The EIS spectra obtained for lipid film before freeze–thaw treatment display features characteristics for defected membranes. Namely, the presence of defects is manifested by a steplike feature on the total impedance curve and pronounced phase minimum observed at frequency ~1 Hz. However, the phase angle minimum of ~60° is quite shallow, which may indicate the heterogeneous distribution of defects within the membrane [42]. Such interpretation is supported by AFM results where numerous randomly distributed pinholes of different sizes were observed. The EIS spectra for lipid film after the freeze–thaw cycle demonstrates that the total impedance is higher, and the phase angle plot displays a plateau within the frequency range of 1–100 Hz, corresponding to a value ~85°. This indicates that the impedance is predominantly determined by capacitance. At lower frequencies, the decrease of the phase angle is observed, which suggests that some defects are still present, but according to the model developed by Valincius and coworkers, it can be concluded that their density is much lower compared with the membrane before freeze-thaw treatment [43]. More quantitative analysis of the electrical properties of lipid films assembled from bicelles can be performed by comparison of the membrane resistance (R_m), which reflects the membrane permeability for water and ions. For this purpose, the EIS data were fitted with the equivalent circuit shown as an inset in Figure 5. This model assumes that the lipid membrane is separated from the electrode surface by a hydrophilic spacer and enables the estimation of membrane resistance. The plot illustrating the changes in values of R_m as a function of the potential is shown in Figure 6.

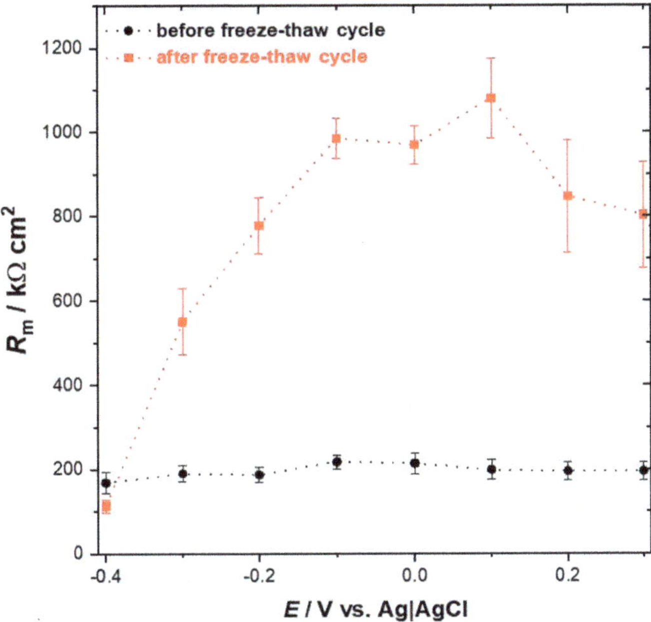

Figure 6. Changes of the membrane resistance (R_m) as a function of the potential applied to the thioglucose-modified Au(111) electrode with lipid assembly before (black) and after (red) one freeze–thaw cycle.

The results obtained for the lipid film before freeze–thaw treatment demonstrate that membrane resistance varies only slightly with potential and its value is within the range of 170–200 kΩ cm^2. Based on AFM data, we may assume that the permeability of this membrane will be strongly affected by the presence of numerous pinholes, which facilitate uncontrolled ion and water flux across the membrane. The latter contributes to the relatively high ionic conductivity of the membrane, which determines the membrane resistance. In contrast, the R_m for lipid film after freeze–thaw treatment varies noticeably with the potential, and its numerical values are significantly higher. Between the potentials of +0.2 V and −0.1 V where the lipid film is most stable, the membrane resistance is ~1.0 MΩ cm^2, which is a manifestation of good insulating properties and low permeability for ions and water molecules. As demonstrated by AFM data, the lipid film undergoes reorganization upon freeze–thaw treatment. The initially formed defected double bilayer is transformed into the virtually defect-free single bilayer with some patches of the double bilayer. The substantially decreased density of defect sites hinders the ionic transport across the membrane, hence its conductivity is decreased. Consequently, the membrane resistance becomes high. More negative polarization of the electrode results in a gradual decrease of the resistance, which might be related to the onset of the electroporation process and/or accumulation of water and ions at the interfacial region typically observed for lipid membranes at the negative potentials [31,41]. Finally, at the potential of −0.4 V, the membrane resistance becomes comparable to that observed for lipid film before the freeze–thaw cycle, indicating that defect density is similar for both systems within that potential range. Comparison of the data indicates clearly that the insulating properties of the lipid film assembled from bicelles are improved upon freeze–thaw treatment. This is also confirmed by the SEIRAS results. Figure 7 shows the SEIRA spectra, which represents the difference in absorbance between the sample spectrum recorded after one freeze–thaw cycle and the reference spectrum recorded before thermal treatment. It demonstrates that freeze–thaw treatment leads to loss of water from the electrode–electrolyte interface, as can be concluded from the presence of the negative v(O–H) and δ(O–H) bands. This effect can be explained by the sealing of the pinholes and reorganization of the lipid film, which leads to the removal of the water molecules from defects. Additionally, the diffusion of water entrapped between the lamellae into the bulk phase may also contribute to the net loss of water, which will be discussed in more detail below. It should also be mentioned that the SEIRA spectrum in Figure 7 does not contain any bands related to v(C–H) and v(C=O) vibrations from lipids. This suggests that the amount of the lipidic material deposited on the electrode surface remains unchanged upon thermal treatment.

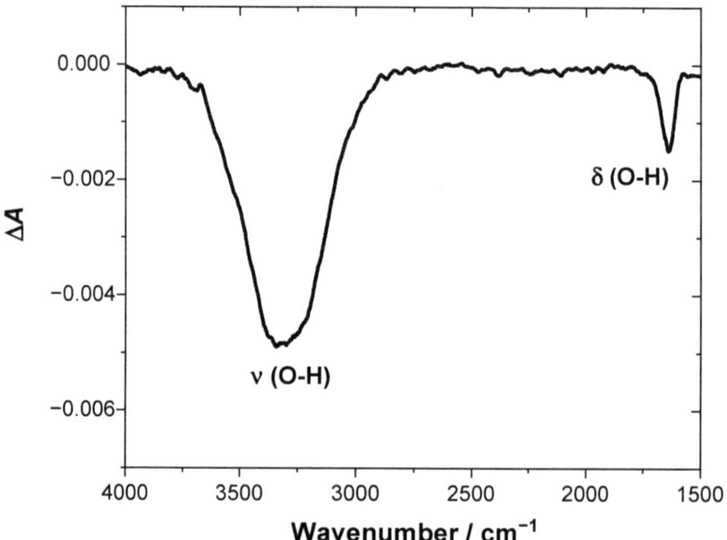

Figure 7. The SEIRA spectra representing the difference in the absorbance between the sample spectrum recorded after one freeze–thaw cycle and the reference spectrum collected before the freeze–thaw cycle. Negative v(O–H) and δ(O–H) bands are indicative of loss of water at the electrode–electrolyte interface.

Based on our experimental observations, we propose the possible scenario of bicelles deposition on thioglucose-modified gold electrode and the transformation of the resulting film into a single bilayer, which is depicted in Figure 8.

Spontaneous deposition of bicellar mixture onto the thioglucose-modified gold electrode results in the formation of the double bilayer with a relatively large number of pinholes. The latter facilitates uncontrolled ion and water transport from the bulk of the solution to the surface of the electrode and contributes to the low resistance of the membrane. Since the double bilayer is a case of a multilamellar system, it means there are two pools of water, i.e., interlamellar water including molecules entrapped between two bilayers as well as water separating the bottom bilayer from the electrode, which can be considered as one pool, and bulk water which can be considered as a second pool. Importantly, these two pools of water have different freezing characteristics. As it was demonstrated for multilamellar lipid vesicles, the interlamellar water can be supercooled to approximately −40 °C before it freezes by homogeneous nucleation [44]. We believe this feature is crucial for the observed reorganization of the membrane. During the freezing cycle, when the temperature is lowered below 4 °C, the ordering of the hydration shell of the top bilayer is expected to decrease and the distance between polar heads is increased [38]. This produces certain asymmetry, which may favor the rearrangement of the molecules to minimize the energy of the system. Further decrease of the temperature causes freezing of the bulk water, while the interlamellar water is liquid, which means that the top membrane experiences the osmotic and mechanical stress, resulting in a diffusion of water molecules to the bulk water phase, and consequently the lamellarity of the film can be decreased similarly as it is observed for the transition from multilamellar to unilamellar vesicles [44,45]. This process is accompanied by the sealing of the pinholes since the lipid material from the top leaflet does not desorb from the electrode surface, but it is redistributed.

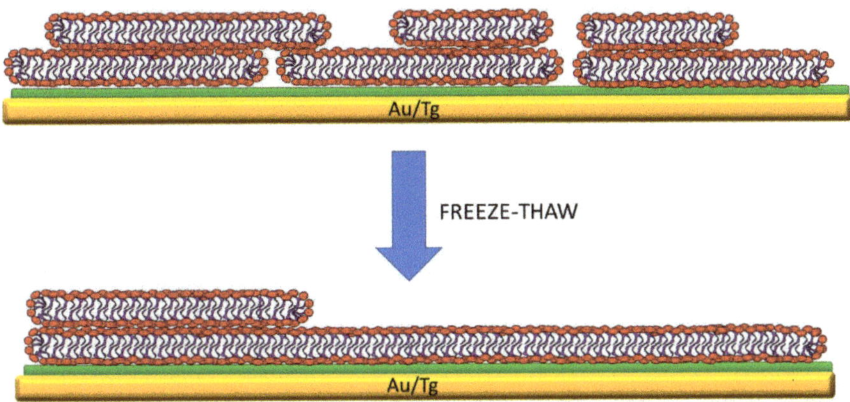

Figure 8. Model illustrating spontaneous deposition of the bicelles on the thioglucose-modified gold electrode and the effect of the freeze–thaw treatment on the structure of the lipid film. Spontaneous deposition of bicelles leads to the formation of a defected double bilayer, but the freeze–thaw treatment causes reorganization of the membrane structure into a defect-free single bilayer with the locally occurring double bilayer.

4. Conclusions

In this work, we have evaluated the properties of the lipid membranes obtained by adsorption of bicelles on thioglucose-modifed gold electrodes. We have demonstrated that adsorption of bicelles occurs by the replacement of the interfacial water. Their spontaneous deposition leads to the formation of a double bilayer structure on the electrode surface. However, the resulting lipid assembly contains numerous defects and pinholes which span either the entire thickness of the film or only the top bilayer. This in turn, affects the electrochemical properties of the membrane, which was found to be permeable for ions and water. Significant improvement in terms of the morphology of the lipid film and its electrochemical characteristics is achieved upon freeze–thaw treatment. The number of pinholes and defect sites is reduced, and the lipid assembly is rearranged to a single bilayer configuration with locally occurring patches of the second bilayer on top of it. Electro-chemical characterization of the lipid membrane after freeze–thaw treatment demonstrated that its permeability for ions and water is significantly reduced. The latter is manifested by the relatively high value of the membrane resistance of ~1 MΩ cm^2. We believe that the approach presented in this work offers an alternative way to obtain stable planar lipid membranes with good insulating properties. In perspective, the benefit of such an approach is that bicelles are known to be a suitable lipid environment for reconstitution and studies of the transmembrane proteins. Hence, more complex cell membrane mimics can be obtained by depositing bicellar mixtures on electrodes.

Supplementary Materials: The following are available online at https://www.mdpi.com/2077-037 5/11/1/11/s1, Figure S1: AFM image of a bare Au(111) electrode collected in an aqueous solution of 0.1 M NaF. The image was taken in PeakForce Tapping mode. Figure S2: AFM image and cross-sectional profile of the ripple phase. The image was collected in an aqueous solution of 0.1 M NaF using PeakForce Tapping mode.

Author Contributions: Conceptualization, D.D. and S.S.; Formal analysis, S.S.; Investigation, D.D.; Methodology, K.S. and S.S.; Resources, K.S.; Writing—original draft, D.D.; Writing—review & editing, S.S. All authors have read and agreed to the published version of the manuscript.

Funding: This research was funded by Polish National Science Centre, Project No. 2016/21/B/ST4/ 02122.

Acknowledgments: The study was carried out at the Biological and Chemical Research Centre, University of Warsaw, established within the project co-financed by European Union from the European Regional Development Fund under the Operational Program Innovative Economy, 2007–2013.

Conflicts of Interest: The authors declare no conflict of interest.

References

1. Sackmann, E. Supported Membranes: Scientific and Practical Applications. *Science* **1996**, *271*, 43–48. [CrossRef] [PubMed]
2. Brian, A.A.; McConnell, H.M. Allogeneic Stimulation of Cytotoxic T Cells by Supported Planar Membranes. *Proc. Natl. Acad. Sci. USA* **1984**, *81*, 6159–6163. [CrossRef] [PubMed]
3. Poursoroush, A.; Sperotto, M.M.; Laradji, M. Phase Behavior of Supported Lipid Bilayers: A Systematic Study by Coarse-Grained Molecular Dynamics Simulations. *J. Chem. Phys.* **2017**, *146*, 154902. [CrossRef] [PubMed]
4. Schneemilch, M.; Quirke, N. Free Energy of Adsorption of Supported Lipid Bilayers from Molecular Dynamics Simulation. *Chem. Phys. Lett.* **2016**, *664*, 199–204. [CrossRef]
5. Castellana, E.T.; Cremer, P.S. Solid Supported Lipid Bilayers: From Biophysical Studies to Sensor Design. *Surf. Sci. Rep.* **2006**, *61*, 429–444. [CrossRef]
6. Becucci, L.; Guidelli, R.; Peggion, C.; Toniolo, C.; Moncelli, M.R. Incorporation of Channel-Forming Peptides in a Hg-Supported Lipid Bilayer. *J. Electroanal. Chem.* **2005**, *576*, 121–128. [CrossRef]
7. Knoll, W.; Köper, I.; Naumann, R.; Sinner, E.-K. Tethered Bimolecular Lipid Membranes—A Novel Model Membrane Platform. *Electrochim. Acta* **2008**, *53*, 6680–6689. [CrossRef]
8. Konarzewska, D.; Juhaniewicz, J.; Güzeloğlu, A.; Sęk, S. Characterization of Planar Biomimetic Lipid Films Composed of Phosphatidylethanolamines and Phosphatidylglycerols from Escherichia Coli. *Biochim. Biophys. Acta Biomembr.* **2017**, *1859*, 475–483. [CrossRef]
9. Matyszewska, D.; Bilewicz, R.; Su, Z.; Abbasi, F.; Leitch, J.J.; Lipkowski, J. PM-IRRAS Studies of DMPC Bilayers Supported on Au(111) Electrodes Modified with Hydrophilic Monolayers of Thioglucose. *Langmuir* **2016**, *32*, 1791–1798. [CrossRef]
10. Su, Z.; Jay Leitch, J.; Abbasi, F.; Faragher, R.J.; Schwan, A.L.; Lipkowski, J. EIS and PM-IRRAS Studies of Alamethicin Ion Channels in a Tethered Lipid Bilayer. *J. Electroanal. Chem.* **2018**, *812*, 213–220. [CrossRef]
11. Richter, R.P.; Brisson, A.R. Following the Formation of Supported Lipid Bilayers on Mica: A Study Combining AFM, QCM-D, and Ellipsometry. *Biophys. J.* **2005**, *88*, 3422–3433. [CrossRef] [PubMed]
12. Richter, R.; Mukhopadhyay, A.; Brisson, A. Pathways of Lipid Vesicle Deposition on Solid Surfaces: A Combined QCM-D and AFM Study. *Biophys. J.* **2003**, *85*, 3035–3047. [CrossRef]
13. Pawłowski, J.; Juhaniewicz, J.; Güzeloğlu, A.; Sek, S. Mechanism of Lipid Vesicles Spreading and Bilayer Formation on a Au(111) Surface. *Langmuir* **2015**, *31*, 11012–11019. [CrossRef] [PubMed]
14. Lipkowski, J. Building Biomimetic Membrane at a Gold Electrode Surface. *Phys. Chem. Chem. Phys.* **2010**, *12*, 13874–13887. [CrossRef] [PubMed]
15. Su, Z.; Leitch, J.J.; Lipkowski, J. Electrode-Supported Biomimetic Membranes: An Electrochemical and Surface Science Approach for Characterizing Biological Cell Membranes. *Curr. Opin. Electrochem.* **2018**, *12*, 60–72. [CrossRef]
16. Valincius, G.; Meškauskas, T.; Ivanauskas, F. Electrochemical Impedance Spectroscopy of Tethered Bilayer Membranes. *Langmuir* **2012**, *28*, 977–990. [CrossRef] [PubMed]
17. Misawa, N.; Osaki, T.; Takeuchi, S. Membrane Protein-Based Biosensors. *J. R. Soc. Interface* **2018**, *15*, 20170952. [CrossRef]
18. Nikoleli, G.-P.; Nikolelis, D.P.; Siontorou, C.G.; Nikolelis, M.-T.; Karapetis, S. The Application of Lipid Membranes in Biosensing. *Membranes* **2018**, *8*, 108. [CrossRef]
19. Vestergaard, M.; Kraft, J.F.; Vosegaard, T.; Thøgersen, L.; Schiøtt, B. Bicelles and Other Membrane Mimics: Comparison of Structure, Properties, and Dynamics from MD Simulations. *J. Phys. Chem. B* **2015**, *119*, 15831–15843. [CrossRef]
20. Li, M.; Morales, H.H.; Katsaras, J.; Kučerka, N.; Yang, Y.; Macdonald, P.M.; Nieh, M.-P. Morphological Characterization of DMPC/CHAPSO Bicellar Mixtures: A Combined SANS and NMR Study. *Langmuir* **2013**, *29*, 15943–15957. [CrossRef]
21. Smrt, S.T.; Draney, A.W.; Singaram, I.; Lorieau, J.L. Structure and Dynamics of Membrane Proteins and Membrane Associated Proteins with Native Bicelles from Eukaryotic Tissues. *Biochemistry* **2017**, *56*, 5318–5327. [CrossRef] [PubMed]
22. Zeineldin, R.; Last, J.A.; Slade, A.L.; Ista, L.K.; Bisong, P.; O'Brien, M.J.; Brueck, S.R.J.; Sasaki, D.Y.; Lopez, G.P. Using Bicellar Mixtures To Form Supported and Suspended Lipid Bilayers on Silicon Chips. *Langmuir* **2006**, *22*, 8163–8168. [CrossRef] [PubMed]
23. Kolahdouzan, K.; Jackman, J.A.; Yoon, B.K.; Kim, M.C.; Johal, M.S.; Cho, N.-J. Optimizing the Formation of Supported Lipid Bilayers from Bicellar Mixtures. *Langmuir* **2017**, *33*, 5052–5064. [CrossRef] [PubMed]

24. Sut, T.N.; Jackman, J.A.; Cho, N.-J. Understanding How Membrane Surface Charge Influences Lipid Bicelle Adsorption onto Oxide Surfaces. *Langmuir* **2019**, *35*, 8436–8444. [CrossRef]
25. Yamada, N.L.; Sferrazza, M.; Fujinami, S. In-Situ Measurement of Phospholipid Nanodisk Adhesion on a Solid Substrate Using Neutron Reflectometry and Atomic Force Microscopy. *Phys. B Condens. Matter* **2018**, *551*, 222–226. [CrossRef]
26. Kycia, A.H.; Wang, J.; Merrill, A.R.; Lipkowski, J. Atomic Force Microscopy Studies of a Floating-Bilayer Lipid Membrane on a Au(111) Surface Modified with a Hydrophilic Monolayer. *Langmuir* **2011**, *27*, 10867–10877. [CrossRef]
27. Kycia, A.H.; Sek, S.; Su, Z.; Merrill, A.R.; Lipkowski, J. Electrochemical and STM Studies of 1-Thio-β-d-Glucose Self-Assembled on a Au(111) Electrode Surface. *Langmuir* **2011**, *27*, 13383–13389. [CrossRef]
28. Ujwal, R.; Bowie, J.U. Crystallizing Membrane Proteins Using Lipidic Bicelles. *Methods* **2011**, *55*, 337–341. [CrossRef]
29. Osawa, M. Topics in Applied Physics. In *Near-Field Optics and Surface Plasmon Polaritons*; Springer: New York, NY, USA, 2002; pp. 163–187.
30. Uchida, T.; Osawa, M.; Lipkowski, J. SEIRAS Studies of Water Structure at the Gold Electrode Surface in the Presence of Supported Lipid Bilayer. *J. Electroanal. Chem.* **2014**, *716*, 112–119. [CrossRef]
31. Juhaniewicz-Dębińska, J.; Konarzewska, D.; Sęk, S. Effect of Interfacial Water on the Nanomechanical Properties of Negatively Charged Floating Bilayers Supported on Gold Electrodes. *Langmuir* **2019**, *35*, 9422–9429. [CrossRef]
32. Su, Z.; Juhaniewicz-Debinska, J.; Sek, S.; Lipkowski, J. Water Structure in the Submembrane Region of a Floating Lipid Bilayer: The Effect of an Ion Channel Formation and the Channel Blocker. *Langmuir* **2019**, *36*, 409–418. [CrossRef] [PubMed]
33. Casal, H.L.; Mantsch, H.H. Polymorphic Phase Behaviour of Phospholipid Membranes Studied by Infrared Spectroscopy. *Biochim. Biophys. Acta* **1984**, *779*, 381–401. [CrossRef]
34. Zawisza, I.; Bin, X.; Lipkowski, J. Potential-Driven Structural Changes in Langmuir−Blodgett DMPC Bilayers Determined by in Situ Spectroelectrochemical PM IRRAS. *Langmuir* **2007**, *23*, 5180–5194. [CrossRef] [PubMed]
35. Juhaniewicz, J.; Sek, S. Atomic Force Microscopy and Electrochemical Studies of Melittin Action on Lipid Bilayers Supported on Gold Electrodes. *Electrochim. Acta* **2015**, *162*, 53–61. [CrossRef]
36. Gál, M.; Sokolová, R.; Naumowicz, M.; Híveš, J.; Krahulec, J. Electrochemical and AFM Study of the Interaction of Recombinant Human Cathelicidin LL-37 with Various Supported Bilayer Lipid Membranes. *J. Electroanal. Chem.* **2018**, *821*, 40–46. [CrossRef]
37. Fukuma, T.; Higgins, M.J.; Jarvis, S.P. Direct Imaging of Individual Intrinsic Hydration Layers on Lipid Bilayers at Ångstrom Resolution. *Biophys. J.* **2007**, *92*, 3603–3609. [CrossRef] [PubMed]
38. Gabbutt, C.; Shen, W.; Seifert, J.; Contera, S. AFM Nanoindentation Reveals Decrease of Elastic Modulus of Lipid Bilayers near Freezing Point of Water. *Sci. Rep.* **2019**, *9*, 19473. [CrossRef]
39. Leidy, C.; Kaasgaard, T.; Crowe, J.H.; Mouritsen, O.G.; Jørgensen, K. Ripples and the Formation of Anisotropic Lipid Domains: Imaging Two-Component Supported Double Bilayers by Atomic Force Microscopy. *Biophys. J.* **2002**, *83*, 2625–2633. [CrossRef]
40. Kaasgaard, T.; Leidy, C.; Crowe, J.H.; Mouritsen, O.G.; Jørgensen, K. Temperature-Controlled Structure and Kinetics of Ripple Phases in One- and Two-Component Supported Lipid Bilayers. *Biophys. J.* **2003**, *85*, 350–360. [CrossRef]
41. Chen, M.; Li, M.; Brosseau, C.L.; Lipkowski, J. AFM Studies of the Effect of Temperature and Electric Field on the Structure of a DMPC−Cholesterol Bilayer Supported on a Au(111) Electrode Surface. *Langmuir* **2009**, *25*, 1028–1037. [CrossRef]
42. Valincius, G.; Mickevicius, M.; Penkauskas, T.; Jankunec, M. Electrochemical Impedance Spectroscopy of Tethered Bilayer Membranes: An Effect of Heterogeneous Distribution of Defects in Membranes. *Electrochim. Acta* **2016**, *222*, 904–913. [CrossRef]
43. Valincius, G.; Mickevicius, M. Chapter Two—Tethered Phospholipid Bilayer Membranes: An Interpretation of the Electrochemical Impedance Response. In *Advances in Planar Lipid Bilayers and Liposomes*; Iglič, A., Kulkarni, C.V., Rappolt, M.B.T.-A., Eds.; Academic Press: New York, NY, USA, 2015; Volume 21, pp. 27–61. [CrossRef]
44. Kaasgaard, T.; Mouritsen, O.G.; Jørgensen, K. Freeze/Thaw Effects on Lipid-Bilayer Vesicles Investigated by Differential Scanning Calorimetry. *Biochim. Biophys. Acta Biomembr.* **2003**, *1615*, 77–83. [CrossRef]
45. Traïkia, M.; Warschawski, D.E.; Recouvreur, M.; Cartaud, J.; Devaux, P.F. Formation of Unilamellar Vesicles by Repetitive Freeze-Thaw Cycles: Characterization by Electron Microscopy and 31P-Nuclear Magnetic Resonance. *Eur. Biophys. J.* **2000**, *29*, 184–195. [CrossRef] [PubMed]

 *membranes*

Article

Electrical Properties of Membrane Phospholipids in Langmuir Monolayers

Anna Chachaj-Brekiesz [1], Jan Kobierski [2], Anita Wnętrzak [1] and Patrycja Dynarowicz-Latka [1,*]

[1] Faculty of Chemistry, Jagiellonian University, Gronostajowa 2, 30-387 Kraków, Poland; anna.chachaj@uj.edu.pl (A.C.-B.); anita.wnetrzak@uj.edu.pl (A.W.)

[2] Department of Pharmaceutical Biophysics, Faculty of Pharmacy, Jagiellonian University Medical College, Medyczna 9, 30-688 Kraków, Poland; jan.kobierski@uj.edu.pl

[*] Correspondence: ucdynaro@cyf-kr.edu.pl

Abstract: Experimental surface pressure (π) and electric surface potential (ΔV) isotherms were measured for membrane lipids, including the following phosphatidylcholines (PCs)—1,2-dipalmitoyl-sn-glycero-3-phosphocholine (DPPC); 1,2-distearoyl-sn-glycero-3-phosphocholine (DSPC); 1,2-diarachidoyl-sn-glycero-3-phosphocholine (DAPC); and 1,2-dioleoyl-sn-glycero-3-phosphocholine (DOPC). In addition, other phospholipids, such as phosphatidylethanolamines (represented by 1,2-dipalmitoyl-sn-glycero-3-phosphoethanolamine (DPPE)) and sphingolipids (represented by N-(hexadecanoyl)-sphing-4-enine-1-phosphocholine (SM)) were also studied. The experimental apparent dipole moments (μ_A^{exp}) of the abovementioned lipids were determined using the Helmholtz equation. The particular contributions to the apparent dipole moments of the investigated molecules connected with their polar ($\mu_\perp^P$) and apolar parts ($\mu_\perp^a$) were theoretically calculated for geometrically optimized systems. Using a three-layer capacitor model, introducing the group's apparent dipole moments (calculated herein) and adopting values from other papers to account for the reorientation of water molecules ($\mu_\perp^w/\varepsilon_w$), as well as the for the local dielectric permittivity in the vicinity of the polar (ε_p) and apolar (ε_a) groups, the apparent dipole moments of the investigated molecules were calculated (μ_A^{calc}). Since the comparison of the two values (experimental and calculated) resulted in large discrepancies, we developed a new methodology that correlates the results from density functional theory (DFT) molecular modeling with experimentally determined values using multiple linear regression. From the fitted model, the following contributions to the apparent dipole moments were determined: $\mu_\perp^w/\varepsilon_w = -1.8 \pm 1.4\ D$; $\varepsilon_p = 10.2 \pm 7.0$ and $\varepsilon_a = 0.95 \pm 0.52$). Local dielectric permittivity in the vicinity of apolar groups (ε_a) is much lower compared to that in the vicinity of polar moieties (ε_p), which is in line with the tendency observed by other authors studying simple molecules with small polar groups. A much higher value for the contributions from the reorientation of water molecules ($\mu_\perp^w/\varepsilon_w$) has been interpreted as resulting from bulky and strongly hydrated polar groups of phospholipids.

Keywords: Langmuir monolayers; electric surface potential; dipole moments; phospholipids

Citation: Chachaj-Brekiesz, A.; Kobierski, J.; Wnętrzak, A.; Dynarowicz-Latka, P. Electrical Properties of Membrane Phospholipids in Langmuir Monolayers. *Membranes* **2021**, *11*, 53. https://doi.org/10.3390/membranes 11010053

Received: 13 December 2020
Accepted: 11 January 2021
Published: 13 January 2021

Publisher's Note: MDPI stays neutral with regard to jurisdictional claims in published maps and institutional affiliations.

1. Introduction

The Langmuir monolayers formed by insoluble amphiphiles at the free water surface have mainly been analyzed using the classical method based on recording surface pressure (π)–area (A) isotherms, which enables researchers to monitor changes of the physical state of film molecules upon compression [1]. Their visualization is possible with microscopic methods like Brewster angle microscopy [2] and fluorescence microscopy [3]. Changes in the electric potential (ΔV), which provide important information on the orientation of molecules at the surface, are performed less frequently. Such measurements complement the characterization of the film's electrical properties (including dipole moments and dielectric permittivity), which play an important role in many intermolecular interactions.

Their analysis provides the basis for an insight into the understanding of biomolecular processes in membranes.

A Langmuir monolayer can be considered as an array of electric dipoles of film-forming molecules situated at the air/water interface. Of particular interest are phospholipid monolayers, which provide a simplified, two-dimensional membrane model that is suitable for studying interactions [4]. Particularly important are those between the polar groups of film molecules and substances dissolved in the subphase (ions and soluble biomolecules). During interactions, the electrical surface potential of the monolayer can be decreased or increased. The measurements of such modifications are of great value as they can be used to screen drugs and check their effectiveness [5–8].

The experimental surface potential changes of a monolayer have usually been interpreted in the terms of so-called effective dipole moments ($\mu_\perp$). In the simplest approach, derived from the parallel plate condenser model [9], ΔV is expressed by the Helmholtz equation:

$$\Delta V = \frac{\mu_\perp}{A \varepsilon \varepsilon_0} \tag{1}$$

where ε is the dielectric permittivity of the film, ε_0 is the dielectric permittivity of the vacuum, $\mu_\perp$. is the normal (to the interface) component of the dipole moment of the film molecule at the interface (note that this is different from the molecular dipole moment of the free molecule) and A is the average area occupied by the molecule at the surface ($A = 1/N$, where N is the total number of molecules at 1 cm^2 of the surface). The above equation applies to un-ionized molecules. For ionized ones, the double layer potential (ψ_0) must be taken into account [9,10]. The main problem in Equation (1) is the unknown value of the permittivity of the film, ε. One of the approaches assumes that $\varepsilon = 1$, either because molecules are considered as isolated entities or because of the lack of a known value [10], however, this can be assumed only for gaseous films. Some authors have claimed that a value $5 < \varepsilon < 10$ should be used [11]. Others suggest that for condensed monolayers, ε can be taken as 2, which is the dielectric permittivity of hydrocarbons [12]. In another approach, the unknown value of ε has been included in the so-called "apparent dipole moment" of a film molecule, $\mu_A = \frac{\mu_\perp}{\varepsilon}$ [13]. μ_A can be easily determined from the experimental values of ΔV as a function of A. Apart from the Helmholtz model, other approaches have been suggested in order to interpret surface potential changes (reviewed in [14,15]). A frequently used model was provided by Demchak and Fort [16], which treats the monolayer as a three-layer capacitor. In this model, the effective dipole moment of a film molecule can be divided into the contributions from the reorientation of water molecules in the monolayer, the polar and apolar part of the film molecule ($\mu_\perp^w$, $\mu_\perp^p$ and $\mu_\perp^a$, respectively), divided by their local dielectric permittivities:

$$\Delta V = \frac{1}{A \varepsilon_0} \left(\frac{\mu_\perp^w}{\varepsilon_w} + \frac{\mu_\perp^p}{\varepsilon_p} + \frac{\mu_\perp^a}{\varepsilon_a} \right) \tag{2}$$

Equation (2) has been used to interpret electric surface potentials for both adsorbed [17] as well as insoluble monolayers [18]. Group dipole moments, $\mu_\perp^p$ and $\mu_\perp^a$ were calculated from bond dipole moments and angles between them, whereas local dielectric permittivities ε_a and ε_p were obtained by solving in pairs equations of type (2) for molecules having the same apolar parts and different polar parts, and vice versa, assuming that the contribution from the reorientation of water molecules ($\mu_\perp^w / \varepsilon_w$) in each pair of equations is the same. Using this procedure for adsorbed films [17] and Langmuir monolayers [18,19] formed by carboxylic acids, alcohols and their derivatives, the following values were obtained: $\varepsilon_p = 4.2$; $\varepsilon_a = 2.4$; $(\mu_\perp^w / \varepsilon_w) = -100 \div -200$ mD; $\varepsilon_p = 6.4$; $\varepsilon_a = 2.8$; $(\mu_\perp^w / \varepsilon_w) = -65$ mD, and $\varepsilon_p = 7.6$; $\varepsilon_a = 4.2$; $(\mu_\perp^w / \varepsilon_w) = 25$ mD, respectively. Upon analyzing surface potential changes for terphenyl derivatives [16], the following values were obtained: $\varepsilon_p = 7.6$; $\varepsilon_a = 5.3$; $(\mu_\perp^w / \varepsilon_w) = 40$ mD. Although there are some differences as regards the values of local dielectric permittivities, ε_p is always higher than ε_a. The greatest discrepancies concern values of ($\mu_\perp^w / \varepsilon_w$), which are generally small, but their sign was determined to be positive or negative.

The history of measuring surface potentials of Langmuir films from membrane lipids is relatively long but there are many discrepancies in the reported values [20–22]. This can result from the applied approaches, laboratory procedures as well as equipment (including types of measuring electrodes used). Although nowadays this technique is routinely applied in order to analyse the electrical properties of surface films [7,8,23,24], in the literature there are results for selected, single molecules only and no systematic studies, especially for biologically important molecules, are available. The main constituent molecules of biomembranes, i.e., phospholipids, are of particular importance. Therefore, the aim of this paper was to provide the characteristics of the electrical properties of the most abundant membrane phospholipids, i.e., phosphatidylcholines (PCs), differing in acyl chain length (1,2-dipalmitoyl-sn-glycero-3-phosphocholine (DPPC); 1,2-distearoyl-sn-glycero-3-phosphocholine (DSPC); 1,2-diarachidoyl-sn-glycero-3-phosphocholin (DAPC) and unsaturation (1,2-dioleoyl-sn-glycero-3-phosphocholine (DOPC). For comparison, other phospholipids, such as phosphatidylethanolamines (represented by 1,2-dipalmitoyl-sn-glycero-3-phosphoethanolamine (DPPE) and sphingolipids (represented by N-(hexadecanoyl)-sphing-4-enine-1-phosphocholine (SM) were also investigated. The chemical structures of the studied phospholipids are shown in Figure 1. All phospholipids selected for our study have net charge zero at pH 7.

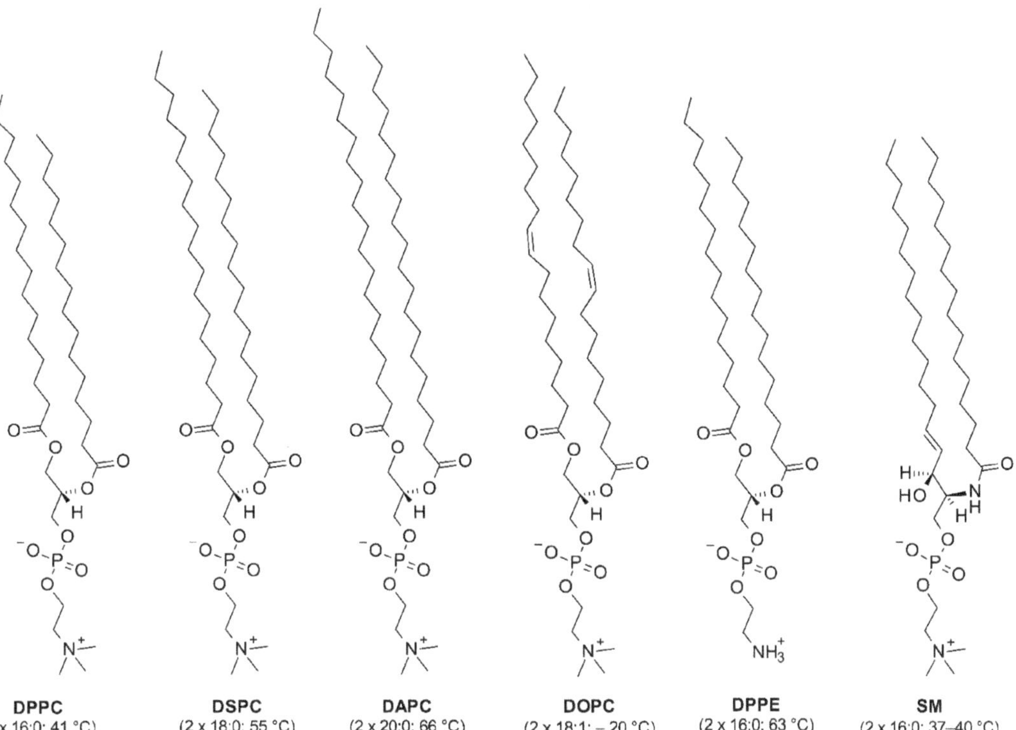

Figure 1. Chemical structures of the studied phospholipids together with acyl chain lengths. Chain melting temperatures of the hydrocarbon chains are given in parentheses. DPPC- 1,2-dipalmitoyl-sn-glycero-3-phosphocholine; DSPC- 1,2-distearoyl-sn-glycero-3-phosphocholine; DAPC -1,2-diarachidoyl-sn-glycero-3-phosphocholine; DOPC- 1,2-dioleoyl-sn-glycero-3-phosphocholine; DPPE- 1,2-dipalmitoyl-sn-glycero-3-phosphoethanolamine; SM- N-(hexadecanoyl)-sphing-4-enine-1-phosphocholine.

The theoretical models used so far for determining the apparent dipole moments and local dielectric permittivities were developed for very simple molecules such as carboxylic acids, alcohols and amines. The use of these models to determine the electrical properties of more complex molecules has not worked well. Therefore, an additional goal of our research was to develop a universal model to be used for molecules of any structure, based on density functional theory (DFT) modeling and multiple linear regression.

2. Materials and Methods

2.1. Surface Potential Measurements

DPPC (1,2-dipalmitoyl-sn-glycero-3-phosphocholine; 16:0 PC), DSPC (1,2-distearoyl-sn-glycero-3-phosphocholine; 18:0 PC), DAPC (1,2-diarachidoyl-sn-glycero-3-phosphocholine; 20:0 PC); DOPC (1,2-dioleoyl-sn-glycero-3-phosphocholine; 18:1 PC), egg SM (containing 86% of N-(hexadecanoyl)-sphing-4-enine-1-phosphocholine; 16:0 SM) and DPPE (1,2-dipalmitoyl-sn-glycero-3-phosphoethanolamine; 16:0 PE) in purities > 99% were supplied by Avanti Polar Lipids. All the investigated phospholipids (except for DOPC) were investigated below their chain melting temperatures (specified in Figure 1). Ethanol (98%) and spectral grade chloroform (stabilized with ethanol) were delivered by Sigma-Aldrich. Spreading solutions for Langmuir experiments were prepared by dissolving each compound in chloroform or chloroform:ethanol (9:1) with a typical concentration of 0.2–0.3 mg·mL^{-1}. In a standard experiment, 50–100 µL of the investigated solution was spread with a microsyringe (precise to ± 2.5 µL). After spreading, the monolayers were left for 10 min for solvent evaporation before starting the compression at a barrier speed of 20 cm^2/min. Deionized ultrapure water from a Millipore system with a resistivity of 18.2 MΩ·cm, pH 7, and a surface tension of 72.8 mN/m was used as a subphase. The subphase temperature (20 °C) was controlled to within 0.1 °C using a thermostat from Julabo. Experiments were carried out with a two-barrier Langmuir 612D NIMA trough (total area 600 cm^2) placed on an antivibration table. Surface pressure was measured with an accuracy of 0.1 mN/m using a Wilhelmy plate made of chromatography paper (Whatman Chr1) as the pressure sensor. Electric surface potential measurements were performed using a Kelvin probe (model KP2, NFT) mounted on a 612D NIMA trough. The vibrating plate was located ca. 2 mm above the water surface while the reference electrode, made from platinum foil, was placed in the water subphase. The surface potential measurements were reproducible to ± 15 mV and ± 2 Å^2 per molecule. Experimental results of surface pressure–area and electric surface potential–area isotherms presented here are representative curves selected from at least three overlapping experiments.

2.2. Theoretical Calculations

The dipole moments and of polar $\mu_{\perp}^{P}$ and apolar $\mu_{\perp}^{a}$ parts of molecules were calculated for previously geometrically optimized systems using the Gaussian 16 software package [25]. Geometry optimization was performed by density functional theory (DFT) modeling. All calculations were performed using the B3LYP functional [26–29] with a 6-311+G(d,p) basis set [30,31] and the D3 version of Grimme's empirical dispersion with the original D3 damping function [32]. Systems were optimized with the default convergence procedures, with no Fermi broadening.

3. Results and Discussion

The results of surface pressure (π) and electric surface potential changes (ΔV) versus area per molecule (A) measurements for the investigated lipids are presented in Figures 1 and 2. Experimental π–A isotherms for all the studied phospholipids are in good agreement with those already published ([33] for DPPC, [34] for DSPC, [35] for DAPC, [36] for DOPC, [37] for DPPE and [38] for SM). Experimental $\Delta V - A$ curves can be characterized by two important parameters: critical area (A_c, which corresponds to the area at which ΔV is triggered off) and maximum value of surface potential (ΔV_{max}). The change in surface potential, observed at A_c occurs at much earlier stages of monolayer compression compared to the surface pressure lift-off area ($A_{lift-off}$) [10]. The onset areas of both surface pressure ($A_{lift-off}$) and surface potential (A_c) occur at the largest areas for the unsaturated phospholipid DOPC. This is due to the steric requirements for molecules with cis unsaturated bonds in their chains, having a coiled conformation, in contrast to saturated all-trans chains. For uncharged compounds, ΔV is approximately constant and close to zero at large areas per molecule. Upon compression, a sharp change (increase or decrease) is observed at the critical area. Changes in the slope of ΔV–A dependence reflect molecular orientation and/or conformational changes in the layer, as ΔV is proportional to the magnitude of the electrostatic field gradient normal to the subphase surface [1]. The maximum of the surface potential, ΔV_{max}, is defined as the value corresponding to the most vertical orientation of molecules, which usually coincides with the maximum condensation of the monolayer, characterized by the maximum compressibility modulus: $C_s^{-1} = -A\left(\frac{d\pi}{dA}\right)$ [39]. Finally, at a collapse point, ΔV becomes approximately constant. The maximum value of the surface potential, ΔV_{max}, is used to calculate the apparent dipole moment μ_A^{exp} of film molecules, using the following equation:

$$\mu_A^{exp} = \frac{\mu_\perp}{\varepsilon} = \varepsilon_0 \cdot A \cdot \Delta V_{max} \tag{3}$$

wherein ε_0 is the vacuum permittivity, ε is the monolayer permittivity (unknown) and A and ΔV_{max} are values extracted from experimental curves corresponding to the maximum compressibility modulus. The values of μ_A^{exp} for all the investigated phospholipids, together with other characteristic parameters for their monolayers, such as $A_{lift-off}$, A_c, ΔV_{max} and A, are summarized in Table 1.

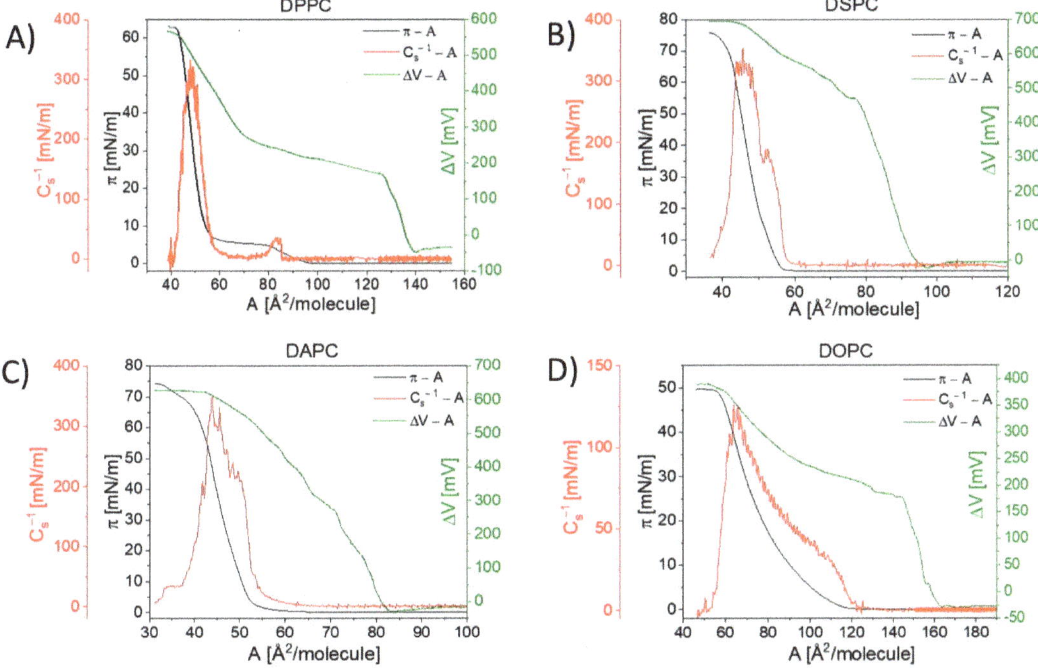

Figure 2. Electric surface potential change and surface pressure isotherms, together with calculated compressional moduli curves for selected phosphatidylcholines: (**A**) DPPC (1,2-dipalmitoyl-sn-glycero-3-phosphocholine), (**B**) DSPC (1,2-distearoyl-sn-glycero-3-phosphocholine), (**C**) DAPC (1,2-diarachidoyl-sn-glycero-3-phosphocholine) and (**D**) DOPC (1,2-dioleoyl-sn-glycero-3-phosphocholine).

Table 1. Selected data read from surface pressure–area (π–A) and electric surface potential–area (ΔV–A) experimental curves measured at 20 °C, together with experimental apparent dipole moments μ_A^{exp} values (uncertainty of $\pm\Delta\mu_A^{exp}$ was obtained by exact differential method)—explanation in the text.

Phospholipid	$A_{lift-off}$ [Å^2/molecule]	A_c [Å^2/molecule]	A [Å^2/molecule]	ΔV_{max} [mV]	μ_A^{exp} [D]	$\pm\Delta\mu_A^{exp}$ [D]
DPPC [1]	96.2	139.0	47.8	509	0.64	0.05
DSPC [2]	57.2	96.2	45.8	684	0.83	0.05
DAPC [3]	58.5	82.6	43.7	614	0.71	0.05
DOPC [4]	116.8	162.3	64.0	358	0.61	0.04
DPPE [5]	62.2	144.2	47.7	573	0.73	0.05
SM [6]	78.3	92.9	42.0	295	0.33	0.03

[1] 1,2-dipalmitoyl-sn-glycero-3-phosphocholine; [2] 1,2-distearoyl-sn-glycero-3-phosphocholine; [3] 1,2-diarachidoyl-sn-glycero-3-phosphocholine; [4] 1,2-dioleoyl-sn-glycero-3-phosphocholine; [5] 1,2-dipalmitoyl-sn-glycero-3-phosphoethanolamine; [6] N-(hexadecanoyl)-sphing-4-enine-1-phosphocholine.

Let us first discuss the surface potential change curves recorded for phosphatidylcholine derivatives (Figure 2), in which the length of the apolar acyl chain or unsaturation degree is varied.

The ΔV–A curve recorded for DPPC shows a characteristic sharp increase starting at the molecular area ca. 139.0 Å^2 from -30 to ca. 168 mV, where there is a visible inflection. Then, the surface potential rise is continued, albeit with a much smaller slope. Starting from ca. 68 Å^2/molecule and 283 mV, the curve slope increases again. A similar sequence is observed in the π–A isotherm and can be attributed to a change in physical surface states in DPPC, i.e., a phase transition from a liquid-expanded (LE) to a liquid-condensed (LC) phase. As is well known, the LC state is more ordered than the LE state due to (i) closer molecular packing (resulting from the increased conformational order of hydrocarbon chains), and (ii) the smaller tilt angle of molecules in the surface layer [40]. The most condensed DPPC monolayer is characterized by a surface potential change equal to 509 mV, which is in agreement with previous studies (527 mV [23], 551 mV [24] and 544 mV [41]). Further elongation of acyl chains to 18 (DSPC) and 20 (DAPC) carbons influence the electrical properties of the formed monolayers. The critical area becomes shifted to smaller values: 96.2 and 82.6 Å^2/molecule for DSPC and DAPC, respectively. This suggests that phosphatidylcholines with longer hydrocarbon chains form more tightly packed and more condensed films compared to the DPPC monolayer. The curve for DSPC is characterized by one inflection appearing at ca. 77 Å^2/molecule and 466 mV, which suggests changes in molecular orientation in the monolayer (understood as a change in molecular angle in respect to the surface) [42]. On the other hand, the ΔV–A isotherm for DAPC shows a gradual rise without noticeable slope changes until the film collapse. The maximum surface potential values are equal to 684 mV and 614 mV for DSPC and DAPC, respectively. The DOPC molecule possesses one double bond in each of the octadecyl chains and can be compared to DSPC, which has hydrophobic chains of the same length, but both are saturated. As can be noticed, the ΔV–A curves of both phospholipids are characterized by the same shape (with one inflection). However, some differences should also be noted. Firstly, the critical area for DOPC is approximately 66 Å^2/molecule larger as compared to DSPC. This suggests lower packing of the DOPC film, which is also confirmed by the values of the compressibility moduli. Secondly, the maximum value of the surface potential ΔV_{max} is equal to 358 mV. Since the surface potential is proportional to the component of the electrostatic field vertical to the surface, it can be concluded that DOPC acyl chains in the monolayer are more disordered in comparison to DSPC. The molecular basis of this issue is due to the presence of double bonds in the cis configuration in DOPC. The configuration of unsaturated bonds affects the conformation of entire hydrocarbon chains, causing disorders (gauche defects, etc.). In contrast, the DSPC hydrocarbon backbone exists mainly in an all-trans zig-zag conformation. To sum up, the participation of hydrocarbon chains in the surface potential change of phosphatidylcholines is mainly steric, related to packing density and order, which is in agreement with previous work [43].

In the next step of our studies, we examined the influence of the interfacial area and polar group modification on surface potential isotherms (Figure 3).

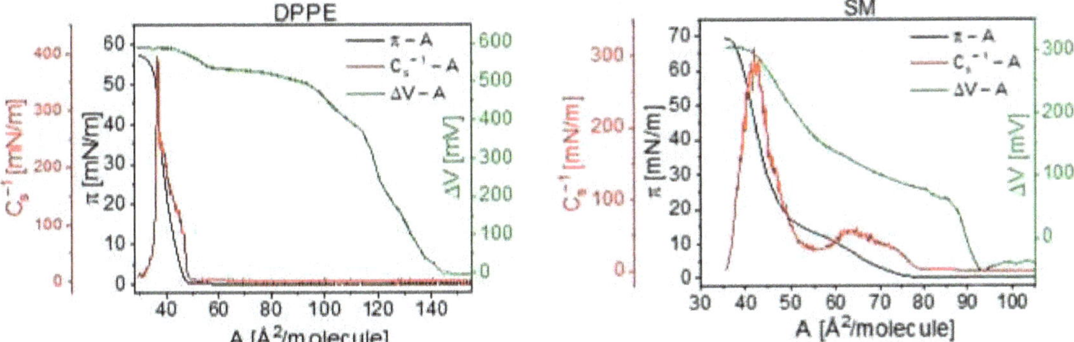

Figure 3. Electric surface potential change and surface pressure isotherms, together with calculated compressional moduli curves for (**A**) DPPE and (**B**) SM.

The SM molecule contains the same polar group as the PCs (the phosphocholine group), however, its structure is different in the region of the sphingosine backbone. The π–A and ΔV–A isotherms measured for SM show an analogical course to DPPC and suggest the existence of stable liquid-expanded and liquid-condensed phases. Despite this fact, the values of the critical area (92.9 Å^2 per molecule) and the maximum of surface potential (295 mV) for SM are significantly lower than for DPPC. Values read from the measured ΔV–A isotherm of SM are in agreement with the literature [24].

In the next part of our studies, we examined the electrical properties of monolayers composed of a representative of phosphatidylethanolamines—DPPE. The ΔV–A curve is characterized by a gradual increase, without any visible inflections, starting from molecular area 144 Å^2 and reaching a maximum value of surface potential equal to 573 mV (in agreement with the literature data, reporting values ranging from 520 mV [18] to 589 mV [23]).

To enrich and interpret the results obtained from our experiments, theoretical DFT calculations of molecular conformations were performed. The values of the total dipole moments of free molecules in a vacuum (μ_{tot}) are compiled in Table 2. To determine the magnitude of the normal component of the electrical dipole moment, μ_z, two different approaches were applied. Firstly, the values μ_z were directly read from the total dipole moment μ_{tot} of the molecule in a vacuum ($\varepsilon = 1$). Additionally, the component of the dipole moment in the plane of the interface (μ_{x-y}) has also been provided. However, it should be emphasized that the calculated dipole moment corresponds to the free molecule, and not to the molecule adsorbed at the interface. Therefore μ_z cannot be directly compared to μ_A^{exp} values obtained from experimental ΔV–A dependencies. In this approach, the molecule is treated as a single entity, without distinguishing its local parts (polar/apolar). As a result, differences in dielectric permittivity between polar and apolar parts are not taken into consideration. In the second approach, we separated the normal component of the dielectric dipole moment of a free molecule into contributions from the polar head ($\mu_\perp^p$) and hydrocarbon chains ($\mu_\perp^a$) individually (see Figure 4). The results of these theoretical calculations (μ_{tot}, μ_{x-y}, μ_z, $\mu_\perp^p$, $\mu_\perp^a$) are collected in Table 2.

In the next step of our calculations, we were interested in relating theoretical and experimental values, using the following equation [16], introducing the values of local dielectric permittivities (ε_p, ε_a) to the theoretically calculated normal dipole moments of the polar and apolar groups ($\mu_\perp^p$, $\mu_\perp^a$) as well as the contribution from the reorientation of water molecules ($\mu_\perp^w / \varepsilon_w$):

$$\mu_A^{calc} = \frac{\mu_\perp^w}{\varepsilon_w} + \frac{\mu_\perp^p}{\varepsilon_p} + \frac{\mu_\perp^a}{\varepsilon_a} \tag{4}$$

In Equation (4), there are three unknown parameters: ($\mu_\perp^w / \varepsilon_w$), ε_p and ε_a. Therefore, initially, we took these values from other works [16,18,19] to see which set of values best fits the value μ_A^{exp} obtained experimentally. The obtained results are presented in Table 3.

Table 2. Values of dipole moments and their contributions, calculated using Gaussian software for entire molecules and their selected parts in vacuum.

Phospholipid	Molecular Dipole Moment in Vacuum (D)			Group Dipole Moment (D)	
	μ_{tot}	μ_{x-y}	μ_z	$\mu_\perp^p$	$\mu_\perp^a$
DPPC	13.07	13.05	0.66	−1.89	2.55
DSPC	13.01	13.00	0.55	−2.16	2.71
DAPC	13.07	13.02	1.06	−1.43	2.49
DOPC	13.04	11.88	5.39	3.40	1.99
DPPE	8.63	8.57	0.94	−1.00	1.94
SM	14.33	9.94	10.33	10.30	0.03

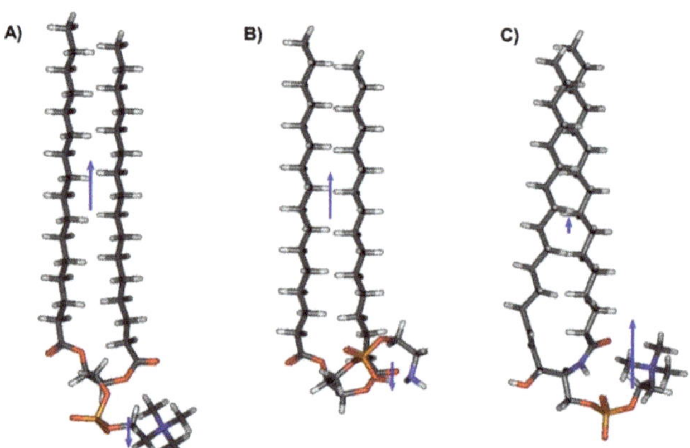

Figure 4. Molecular structure of selected phospholipids: DPPC (**A**), DPPE (**B**) and SM (**C**) with the directions of normal components of electrical dipole moment visualized separately for the polar group ($\mu_\perp^p$) and hydrocarbon chains ($\mu_\perp^a$). Carbon atoms are visualized in dark gray, hydrogen in bright gray, oxygen in red, nitrogen in blue and phosphorus in orange.

As can be seen, the results obtained using the literature values of parameters do not agree with the experimentally determined μ_A^{exp}. Therefore, in the second step, we developed a new methodology. For a series of phosphatidylcholines DPPC, DSPC, DAPC and DOPC, multiple linear regression was used to find a model describing the relationship between the normal components of the calculated dipole moments of the polar and apolar parts of molecules and their experimental values $\left(\mu_A^{exp} \right)$. The equation of the fitted model has the following form:

$$\mu_A^{calc(4)} = -1.8 + 0.098 \cdot \mu_\perp^p + 1.05 \cdot \mu_\perp^a \tag{5}$$

which means that the determined parameters are equal to: $\mu_\perp^w / \varepsilon_w = -1.8 \pm 1.4$ D; $\varepsilon_p = 10.2 \pm 7.0$ and $\varepsilon_a = 0.95 \pm 0.52$. The predicted parameters show good agreement with the observed parameters, as shown in Figure 5. The R-squared statistic indicates that the fitted model explains 86% of the variability in μ_A.

Table 3. Values of apparent dipole moments calculated from Equation (4) using the set of values of $(\mu_\perp^w / \varepsilon_w)$, ε_p and ε_a equal to (1) 0.04; 7.6 and 5.3 (according to [16]); (2) -0.065; 6.4 and 2.8 (according to [18]); (3) 0.04; 7.6; 4.2 (according to [19]); (4) -1.8; 10.2; 0.95 (according to our model).

Phospholipid	Apparent Dipole Moment (D)			
	$\mu_A^{calc(1)}$	$\mu_A^{calc(2)}$	$\mu_A^{calc(3)}$	$\mu_A^{calc(4)}$
DPPC	0.32	0.60	0.44	0.68
DSPC	0.27	0.57	0.40	0.83
DAPC	0.27	0.55	0.40	0.67
DOPC	0.86	1.18	0.96	0.61
DPPE	0.27	0.47	0.37	0.13
SM	1.40	1.55	1.40	-0.78

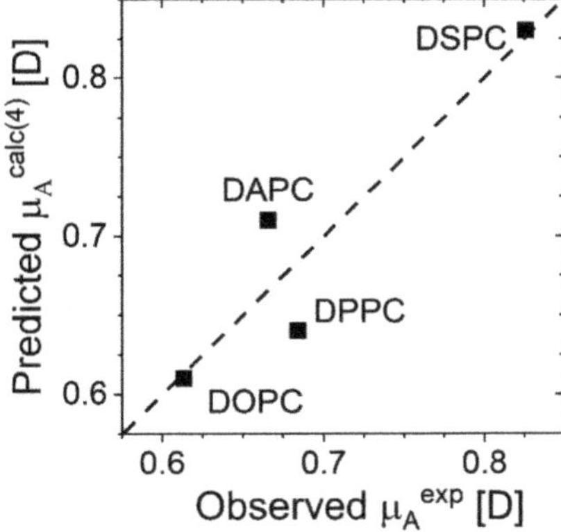

Figure 5. Predicted $\mu_A^{calc(4)}$ versus experimental μ_A^{exp} values for the studied phosphatidylcholines with the line of identity.

As can be seen, the second value (ε_a) is about ten times smaller compared to ε_p, which indicates that the contribution of non-polar groups to the apparent dipole moment is more significant. A similar relationship was found by other authors [16,18,19], however, this effect was not so pronounced. This can result from the fact that literature models were based on homological series of compounds with small, simple polar groups (such as $-COOH$, $-OH$, $-NH_2$), whereas our approach involves bulky, zwitterionic systems. The obtained contribution from the reorientation of water molecules $\mu_\perp^w / \varepsilon_w$ is high, which can be explained by the ability of phosphatidylcholines to form hydrogen bonds with surrounding water dipoles. It is known that the hydration of the polar PC head groups is very high (even 11 water molecules per DPPC [44]), which may result in the formation of an organized water "quasi ice" lattice in the vicinity of a phosphocholine moiety [45]. The model developed in this paper for phosphatidylcholines cannot be applied to other phospholipids due to different hydration of polar groups (i.e., hydration for PC was determined to be 11.3 water molecules per lipid, whereas for PE it is 6.6). Although SM possesses the same polar group as PC, the presence of the hydroxyl group in the interfacial region may affect its hydration [46].

4. Conclusions

In this work, a systematic study on the surface potentials of Langmuir monolayers formed by the most abundant mammalian membrane phospholipids was performed to determine the values of their apparent dipole moments and to propose an improved protocol for estimating individual contributions to a three-layer capacitor model. Our methodology correlates the results from DFT molecular modeling with experimentally determined values using multiple linear regression. It can be applied to the study of other phosphatidylcholine derivatives (with similar hydration) with different hydrocarbon chain lengths and saturations. To be able to determine the parameters ($\mu_\perp^w / \varepsilon_w$), ε_p and ε_a for other kinds of phospholipids, a similar approach should be carried out for their homologous series.

Author Contributions: Conceptualization P.D.-L. and J.K.; investigation, A.C.-B. and J.K.; data curation and visualization, J.K., A.C.-B. and A.W.; manuscript writing P.D.-L., A.C.-B. and A.W. All authors have read and agreed to the published version of the manuscript.

Funding: This research received no external funding.

Institutional Review Board Statement: Not applicable.

Informed Consent Statement: Not applicable.

Acknowledgments: This research was supported in part by PL–Grid Infrastructure.

Conflicts of Interest: The authors declare no conflict of interest regarding the publication of this article.

References

1. Barnes, G.; Gentle, I. *Interfacial Science: An Introduction*, 2nd ed.; Oxford University Press: Oxford, UK, 2011; ISBN 9780199571185.
2. Daear, W.; Mahadeo, M.; Prenner, E.J. Applications of Brewster angle microscopy from biological materials to biological systems. *Biochim. Biophys. Acta Biomembr.* **2017**, *1859*, 1749–1766. [CrossRef] [PubMed]
3. Stottrup, B.L.; Nguyen, A.H.; Tüzel, E. Taking another look with fluorescence microscopy: Image processing techniques in Langmuir monolayers for the twenty-first century. *Biochim. Biophys. Acta Biomembr.* **2010**, *1798*, 1289–1300. [CrossRef] [PubMed]
4. Nobre, T.M.; Pavinatto, F.J.; Caseli, L.; Barros-Timmons, A.; Dynarowicz-Łatka, P.; Oliveira, O.N. Interactions of bioactive molecules & nanomaterials with Langmuir monolayers as cell membrane models. *Thin Solid Films* **2015**, *593*, 158–188.
5. Jaroque, G.N.; Sartorelli, P.; Caseli, L. The effect of the monocyclic monoterpene tertiary alcohol γ-terpineol on biointerfaces containing cholesterol. *Chem. Phys. Lipids* **2020**, *230*, 104915. [CrossRef]
6. Ceridório, L.F.; Caseli, L.; Oliveira, O.N. Chondroitin sulfate interacts mainly with headgroups in phospholipid monolayers. *Colloids Surf. B. Biointerfaces* **2016**, *141*, 595–601. [CrossRef]
7. Nakahara, H.; Hagimori, M.; Mukai, T.; Shibata, O. Monolayers of a tetrazine-containing gemini amphiphile: Interplays with biomembrane lipids. *Colloids Surf. B. Biointerfaces* **2018**, *164*, 1–10. [CrossRef]
8. Nakahara, H.; Hagimori, M.; Mukai, T.; Shibata, O. Interactions of a Tetrazine Derivative with Biomembrane Constituents: A Langmuir Monolayer Study. *Langmuir* **2016**, *32*, 6591–6599. [CrossRef]
9. Dynarowicz, P. Recent developments in the modeling of the monolayers structure at the water/air interface. *Adv. Colloid Interface Sci.* **1993**, *45*, 215–241. [CrossRef]
10. Oliveira, O.N.; Bonardi, C. The Surface Potential of Langmuir Monolayers Revisited. *Langmuir* **1997**, *13*, 5920–5924. [CrossRef]
11. Adam, N.K.; Danielli, J.F.; Harding J., B. The structure of surface films XXI—Surface potentials of dibasic esters, alcohols, aldoximes, and ketones. *Proc. Math. Phys. Eng. Sci.* **1934**, *147*, 491–499.
12. Schuhmann, D. Electrical properties of adsorbed or spread films: The effective value of permittivities in the Helmholtz equation (plane distribution of point dipoles). *J. Colloid Interface Sci.* **1990**, *134*, 152–160. [CrossRef]
13. Broniatowski, M.; Macho, I.S.; Miñones, J.; Dynarowicz-Łatka, P. Langmuir monolayers characteristic of (Perfluorodecyl)-alkanes. *J. Phys. Chem. B.* **2004**, *108*, 13403–13411. [CrossRef]
14. Shibata, O.; Krafft, M.P. Mixed Langmuir monolayers made from single-chain perfluoroalkylated amphiphiles. *Langmuir* **2000**, *16*, 10281–10286. [CrossRef]
15. Oliveira, O.N.; Taylor, D.M.; Morgan, H. Modelling the surface potential-area dependence of a stearic acid monolayer. *Thin Solid Films* **1992**, *210–211*, 76–78. [CrossRef]
16. Demchak, R.J.; Fort, T. Surface dipole moments of close-packed un-ionized monolayers at the air-water interface. *J. Colloid Interface Sci.* **1974**, *46*, 191–202. [CrossRef]
17. Dynarowicz, P.; Paluch, M. Electrical properties of some adsorbed films at the water-air interface. *J. Colloid Interface Sci.* **1985**, *107*, 75–80. [CrossRef]
18. Oliveira, O.N.; Taylor, D.M.; Lewis, T.J.; Salvagno, S.; Stirling, C.J.M. Estimation of group dipole moments from surface potential measurements on Langmuir monolayers. *J. Chem. Soc. Faraday Trans. 1 Phys. Chem. Condens. Phases* **1989**, *85*, 1009–1018. [CrossRef]
19. Petrov, J.G.; Polymeropoulos, E.E.; Möhwald, H. Three-capacitor model for surface potential of insoluble monolayers. *J. Phys. Chem.* **1996**, *100*, 9860–9869. [CrossRef]

20. Mozaffary, H. On the sign and origin of the surface potential of phospholipid monolayers. *Chem. Phys. Lipids* **1991**, *59*, 39–47. [CrossRef]
21. Beitinger, H.; Vogel, V.; Möbius, D.; Rahmann, H. Surface potentials and electric dipole moments of ganglioside and phospholipid monolayers: Contribution of the polar headgroup at the water/lipid interface. *Biochim. Biophys. Acta Biomembr.* **1989**, *984*, 293–300. [CrossRef]
22. Smaby, J.M.; Brockman, H.L. Surface dipole moments of lipids at the argon-water interface. Similarities among glycerol-ester-based lipids. *Biophys. J.* **1990**, *58*, 195–204. [CrossRef]
23. Nakahara, H.; Nakamura, S.; Nakamura, K.; Inagaki, M.; Aso, M.; Higuchi, R.; Shibata, O. Cerebroside Langmuir monolayers originated from the echinoderms: I. Binary systems of cerebrosides and phospholipids. *Colloids Surf. B. Biointerfaces* **2005**, *42*, 157–174. [CrossRef] [PubMed]
24. Sakamoto, S.; Nakahara, H.; Shibata, O. Miscibility behavior of sphingomyelin with phytosterol derivatives by a Langmuir monolayer approach. *J. Oleo Sci.* **2013**, *62*, 809–824. [CrossRef] [PubMed]
25. Frisch, M.J.; Trucks, G.W.; Schlegel, H.B.; Scuseria, G.E.; Robb, M.A.; Cheeseman, J.R.; Scalmani, G.; Barone, V.; Petersson, G.A.; Nakatsuji, H.; et al. *Gaussian 16, Revision B.01*; Gaussian Inc.: Wallingford, CT, USA, 2016.
26. Becke, A.D. Density-functional thermochemistry. III. The role of exact exchange. *J. Chem. Phys.* **1993**, *98*, 5648–5652. [CrossRef]
27. Lee, C.; Yang, W.; Parr, R.G. Development of the Colle-Salvetti correlation-energy formula into a functional of the electron density. *Phys. Rev. B.* **1988**, *37*, 785–789. [CrossRef]
28. Vosko, S.H.; Wilk, L.; Nusair, M. Accurate spin-dependent electron liquid correlation energies for local spin density calculations: A critical analysis. *Can. J. Phys.* **1980**, *58*, 1200–1211. [CrossRef]
29. Stephens, P.J.; Devlin, F.J.; Chabalowski, C.F.; Frisch, M.J. Ab Initio Calculation of Vibrational Absorption and Circular Dichroism Spectra Using Density Functional Force Fields. *J. Phys. Chem.* **1994**, *98*, 11623–11627. [CrossRef]
30. McLean, A.D.; Chandler, G.S. Contracted Gaussian basis sets for molecular calculations. I. Second row atoms, Z = 11–18. *J. Chem. Phys.* **1980**, *72*, 5639–5648. [CrossRef]
31. Krishnan, R.; Binkley, J.S.; Seeger, R.; Pople, J.A. Self-consistent molecular orbital methods. XX. A basis set for correlated wave functions. *J. Chem. Phys.* **1980**, *72*, 650–654. [CrossRef]
32. Grimme, S.; Antony, J.; Ehrlich, S.; Krieg, H. A consistent and accurate ab initio parametrization of density functional dispersion correction (DFT-D) for the 94 elements H-Pu. *J. Chem. Phys.* **2010**, *132*, 154104. [CrossRef]
33. Stefaniu, C.; Brezesinski, G.; Möhwald, H. Polymer-capped magnetite nanoparticles change the 2D structure of DPPC model membranes. *Soft Matter* **2012**, *8*, 7952–7959. [CrossRef]
34. Kilic, S.; Bolukcu, E.S. Phase behavior of DSPC/PEG40St mixtures at higher emulsifier contents. *Colloids Surf. B. Biointerfaces* **2018**, *171*, 368–376. [CrossRef]
35. Borden, M.A.; Pu, G.; Runner, G.J.; Longo, M.L. Surface phase behavior and microstructure of lipid/PEG-emulsifier monolayer-coated microbubbles. *Colloids Surf. B Biointerfaces* **2004**, *35*, 209–223. [CrossRef] [PubMed]
36. Raju, R.; Torrent-Burgués, J.; Bryant, G. Interactions of cryoprotective agents with phospholipid membranes—A Langmuir monolayer study. *Chem. Phys. Lipids* **2020**, *231*, 104949. [CrossRef] [PubMed]
37. Korchowiec, B.; Paluch, M.; Corvis, Y.; Rogalska, E. A Langmuir film approach to elucidating interactions in lipid membranes: 1,2-dipalmitoyl-sn-glycero-3-phosphoethanolamine/cholesterol/metal cation systems. *Chem. Phys. Lipids* **2006**, *144*, 127–136. [CrossRef] [PubMed]
38. Wnętrzak, A.; Makyła-Juzak, K.; Filiczkowska, A.; Kulig, W.; Dynarowicz-Latka, P. Oxysterols Versus Cholesterol in Model Neuronal Membrane. I. The Case of 7-Ketocholesterol. The Langmuir Monolayer Study. *J. Membr. Biol.* **2017**, *250*, 553–564. [CrossRef] [PubMed]
39. Davies, J.T.; Rideal, E.K. *Interfacial phenomena*; Academic Press: Cambridge, MA, USA, 1963; ISBN 9780123956095.
40. Gragson, D.E.; Beaman, D.; Porter, R. Using compression isotherms of phospholipid monolayers to explore critical phenomena. A biophysical chemistry experiment. *J. Chem. Educ.* **2008**, *85*, 272–275. [CrossRef]
41. Taylor, D.M.; De Oliveira, O.N.; Morgan, H. Models for interpreting surface potential measurements and their application to phospholipid monolayers. *J. Colloid Interface Sci.* **1990**, *139*, 508–518. [CrossRef]
42. Sanchez, J.; Badia, A. Spatial variation in the molecular tilt orientational order within the solid domains of phase-separated, mixed dialkylphosphatidylcholine monolayers. *Chem. Phys. Lipids* **2008**, *152*, 24–37. [CrossRef]
43. Peterson, U.; Mannock, D.A.; Lewis, R.N.A.H.; Pohl, P.; McElhaney, R.N.; Pohl, E.E. Origin of membrane dipole potential: Contribution of the phospholipid fatty acid chains. *Chem. Phys. Lipids* **2002**, *117*, 19–27. [CrossRef]
44. Pasenkiewicz-Gierula, M.; Baczynski, K.; Markiewicz, M.; Murzyn, K. Computer modelling studies of the bilayer/water interface. *Biochim. Biophys. Acta Biomembr.* **2016**, *1858*, 2305–2321. [CrossRef] [PubMed]
45. Disalvo, E.A.; Lairion, F.; Martini, F.; Almaleck, H.; Diaz, S.; Gordillo, G. Water in biological membranes at interfaces: Does it play a functional role? *An. des la Asoc. Quim. Argentina* **2004**, *92*, 1–22.
46. Kodama, M.; Abe, M.; Kawasaki, Y.; Hayashi, K.; Ohira, S.; Nozaki, H.; Katagiri, C.; Inoue, K.; Takahashi, H. Estimation of interlamellar water molecules in sphingomyelin bilayer systems studied by DSC and X-ray diffraction. *Thermochim. Acta* **2004**, *416*, 105–111.

Article

Study of Resveratrol's Interaction with Planar Lipid Models: Insights into Its Location in Lipid Bilayers

Daniela Meleleo 

Department of Biosciences, Biotechnologies and Biopharmaceutics, University of Bari "Aldo Moro", via E. Orabona 4, 70126 Bari, Italy; danielaaddolorata.meleleo@uniba.it; Tel.: +39-080-5442775

Abstract: Resveratrol, a polyphenolic molecule found in edible fruits and vegetables, shows a wide range of beneficial effects on human health, including anti-microbial, anti-inflammatory, anti-cancer, and anti-aging properties. Due to its poor water solubility and high liposome-water partition coefficient, the biomembrane seems to be the main target of resveratrol, although the mode of interaction with membrane lipids and its location within the cell membrane are still unclear. In this study, using electrophysiological measurements, we study the interaction of resveratrol with planar lipid membranes (PLMs) of different composition. We found that resveratrol incorporates into palmitoyl-oleoyl-phosphatidylcholine (POPC) and POPC:Ch PLMs and forms conductive units unlike those found in dioleoyl-phosphatidylserine (DOPS):dioleoyl-phosphatidylethanolamine (DOPE) PLMs. The variation of the biophysical parameters of PLMs in the presence of resveratrol provides information on its location within a lipid double layer, thus contributing to an understanding of its mechanism of action.

Keywords: resveratrol; planar lipid membrane; cholesterol; channel-like event; capacitance

Citation: Meleleo, D. Study of Resveratrol's Interaction with Planar Lipid Models: Insights into Its Location in Lipid Bilayers. *Membranes* **2021**, *11*, 132. https://doi.org/10.3390/membranes11020132

Academic Editor: Monika Naumowicz

Received: 18 January 2021
Accepted: 10 February 2021
Published: 14 February 2021

Publisher's Note: MDPI stays neutral with regard to jurisdictional claims in published maps and institutional affiliations.

1. Introduction

Trans 3, 4′, 5-trihydroxystilbene, also known as *Trans*-resveratrol (Figure 1), is found in edible fruits and vegetables, particularly in grapes, blueberries, blackberries, and peanuts.

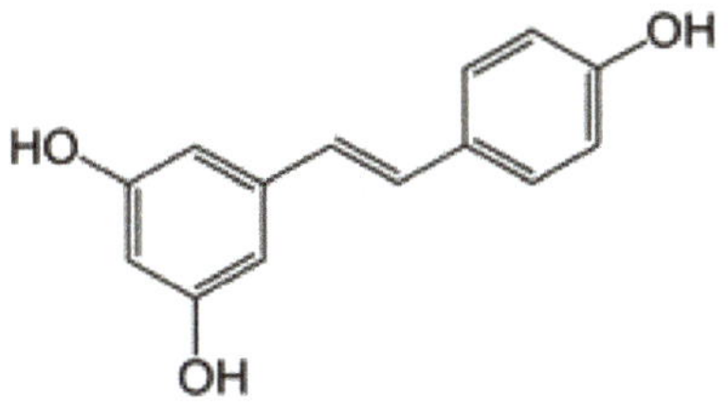

Figure 1. Structure of *Trans*-resveratrol.

Resveratrol is also present in red wine, giving rise to the so-called "French paradox" by which, despite the fat-rich diets, mortality from coronary heart disease is lower in France than in other countries due to the inhabitants' moderate consumption of red wine [1].

Besides its cardio-protective effects, resveratrol has shown a wide range of beneficial effects on human health, including anti-microbial [2], anti-inflammatory [3,4], anti-cancer [5,6], and anti-aging [7,8] properties. Some studies have shown that this polyphenol has a preventive effect on Alzheimer's disease and dementia, due to its anti-inflammatory properties [9–15].

Numerous studies have been carried out in order to gain an understanding of the mechanisms underlying the beneficial effects of resveratrol [16]. The results of studies carried out by Yashiro et al. [17] and Voloshyna et al. [18] show that its cardio-protective effect may be due to resveratrol's capacity to act as a modulator of lipoprotein metabolism. Indeed, resveratrol inhibits the oxidation of low-density lipoproteins and platelet aggregation, while increasing high-density lipoproteins [19]. It is known to be a scavenger of free radicals [20]. Several studies have shown that, when phospholipase A2 promotes the release of cytokines and NADPH oxidases that cause cell inflammation, resveratrol has proven to act as a good cell protective agent [21–24].

Although resveratrol's beneficial effects on human health have been accepted, its molecular mechanism remains unclear. The biomembrane seems to be the main target of resveratrol, although its mode of interaction with membrane lipids and its location within the cell membrane are subject to debate. Resveratrol's affinity for the lipid matrix is indicated by its poor water solubility (0.03 g/mL) and high partition coefficient (3.07 and 3.11 in 1,2-dipalmitoyl-*sn*-glycero-3-phosphocholine (DPPC) and 1,2-distearoyl-*sn*-glycero-3-phosphocholine (DSPC) liposomal membranes, respectively) [25].

Numerous studies have examined the interaction and location of resveratrol within the membrane utilizing different techniques and lipid models given that the cell membrane is a complex system. The results of these studies are not unique. Indeed, it has been either concluded that resveratrol is located in the hydrophilic headgroup region or, on the contrary, in the hydrophobic core.

The results of a study carried out by de Ghellinck and colleagues indicate that resveratrol accumulates between the lipid headgroups, causing conformational changes in the tilt angle of the lipid headgroups to a more upright orientation [26]. Incorporation and location of resveratrol in the hydrophilic headgroup region seem to depend on the lipid composition of the membrane model. The study carried out by Han and colleagues shows that, in 1,2-dioleoyl-sn-glycero-3-phosphatidylcholine (DOPC) liposomes, resveratrol could penetrate into the lipid membrane, locating mostly in the headgroup region rather than in the deeper region of the lipid double layer. In DOPC/sphingomyelin/cholesterol or sphingomyelin/cholesterol liposomes, resveratrol associates with the liposome surface without penetrating into the headgroup region. This different behavior shown by resveratrol may be due to a different packing of the lipid acyl carbon chains. In DOPC liposomes, the acyl carbon chains are less packed than in liposomes containing sphingomyelin and cholesterol [27]. Similar results have also been obtained by Cardia and colleagues, who studied the interaction of resveratrol with soybean phosphatidylcholine (P90G) liposomes by ^{1}H NMR spectroscopy. Resveratrol is located in the region of the phosphocholine headgroup. Quantitative data on the incorporation of resveratrol at both the liposome preparation stage and by preformed liposomes suggest that the amount of resveratrol incorporated into P90G liposomes was about 20 mM, which corresponds to a 150-fold increase with respect to the solubility of resveratrol in water. These experimental data show that resveratrol diffuses through lipid bilayers [28]. Using liposomes made from saturated phosphatidylcholine (DPPC) and cholesterol (CHOL) or positively charged derivates of cholesterol (DC-CHOL), Bonechi et al. found that resveratrol associates with the surface of liposome containing CHOL, as found by Han et al. [27], while the polyphenolic compound inserts more deeply into cationic liposomes [29].

On the other hand, some authors state that resveratrol is located in the hydrophobic core. The results of a study carried out by Balanč and colleagues indicate that resveratrol is incorporated in the inner part of the liposome membrane [30]. In agreement with the results obtained by Balanč are those presented by Fabris and colleagues who, using phosphatidylcholine liposomes of various chain lengths, show that the hydroxyl groups of resveratrol are located in the lipid region of the bilayers close to the double bonds of polyunsaturated fatty acids and that the longer the acyl chains, the more stable and less perturbable the bilayers [25]. Similar results have also been obtained by Brittes and colleagues [1]. In addition, Neves and colleagues show that resveratrol induces phase separation and formation of liquid-ordered domains in liposomes made of egg L-α phosphatidylcholine, cholesterol and sphingomyelin. Besides, this polyphenol stabilizes the membrane bilayers of the liposome locating in a deeper region of the membrane and adopting a vertical position in the nonpolar region of the membrane, with two hydroxyl groups in the interfaces [31,32].

The penetration of a drug into the lipid bilayer is closely related to its antioxidant activity [33,34] because it affects membrane fluidity and polarity. Indeed, numerous studies indicate that increases or decreases in membrane fluidity are responsible for the antioxidant effects of many drugs. Antioxidant drug molecules can trap free radicals or hamper their diffusion by fluidifying [1,25,35] or stiffening [36–38] the membrane, respectively. Resveratrol is able to fluidize and permeate the membrane, suggesting that it is a good antioxidant agent. Besides, the ability of resveratrol to penetrate into the membrane may depend on differently packed lipid, fluid or gel phases [1]. This behavior of resveratrol could explain the controversies regarding its location within the membrane.

In this paper, we report a systematic investigation of resveratrol incorporation into planar lipid membranes (PLMs) and formation of channel-like events using electrophysiological measurements, a new technical approach for the study of resveratrol's membrane activity, aiming to contribute to understanding the action of resveratrol on membranes, by varying PLM electrical parameters (capacitance and conductance) when the polyphenolic compound interacts with it. PLMs of different composition were used for the following reasons: they are a convenient tool to use for protein or drug incorporation, as they are a less complex membrane model system than plasma membranes; they complement studies performed with liposomes.

Incorporation into membranes and channel-like event formations are a test of resveratrol's affinity for the lipid core of PLMs, and, albeit indirectly, they can provide information on its location within the lipid double layer.

2. Materials and Methods

2.1. Reagents and Equipment

Salts and other basic chemicals were bought from Sigma (Munich, FRG, analytical grade). The phospholipids were purchased as follows: palmitoyl-oleoyl-phosphatidylcholine (POPC) from Avanti Polar Lipids, (Alabaster, AL, USA); dioleoyl-phosphatidylserine (DOPS); dioleoyl-phosphatidylethanolamine (DOPE); Cholesterol (Ch) from Sigma (Munich, FRG), resveratrol was from Farmalabor® S.r.l. (Canosa of Puglia, Italy). Figure 2 reports the molecular structures of the lipids used in this study. The volumes (Vl) of phospholipids are similar (Vl = 1231.2/1228/1254 Å^3 POPC/DOPE/DOPS, respectively) [39–41], while their shapes are different. Indeed, POPC and DOPS are classified as cylindrical lipids, and DOPE as a conical lipid [42].

Single-channel measurements: the membrane current was monitored with an oscilloscope and recorded on a chart recorder for further analysis by hand. A Teflon chamber made up of two aqueous compartments connected by a small circular hole with a diameter of 300 μm was used. The *cis* and *trans* chambers were connected to the amplifier head stage by Ag/AgCl electrodes in series with a voltage source and a highly sensitive current amplifier (OPA 129). The single-channel instrumentation had a time resolution of 1–10 ms depending on the magnitude of the single-channel conductance. Figure 3 reports the experimental set-up used to study the interaction of resveratrol with PLMs. The polarity of the voltage was defined according to the side where resveratrol was added (the *cis* side). A trans-negative potential (indicated by a minus sign) means that a negative potential was applied to the *trans* side, the compartment opposite the one where resveratrol was added. These experiments are considered to be a sensitive tool for the study of channel-forming substances [43,44].

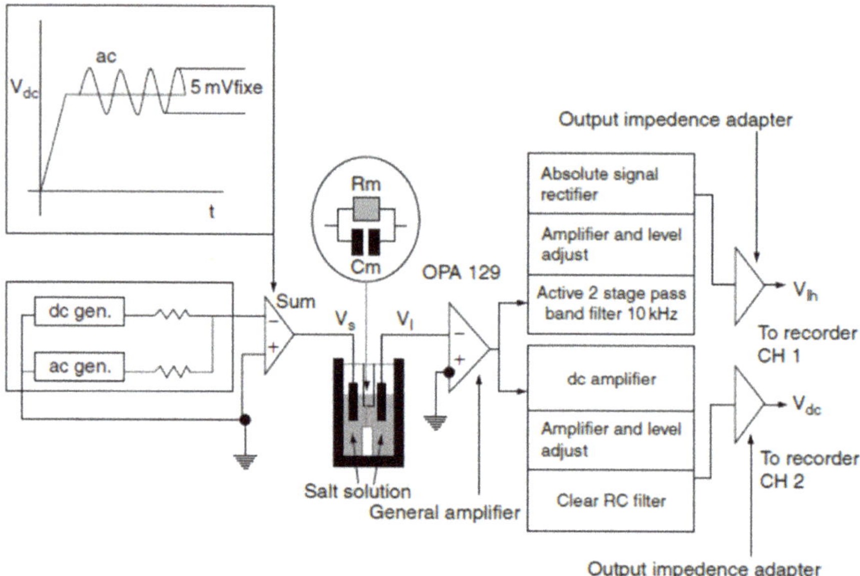

Figure 2. Molecular structures of (**a**) palmitoyl-oleoyl-phosphatidylcholine (POPC), (**b**) dioleoyl-phosphatidylethanolamine (DOPE), (**c**) dioleoyl-phosphatidylserine (DOPS), (**d**) and Cholesterol (Ch).

Figure 3. Experimental set-up during investigations into resveratrol channel-like incorporation.

2.2. Experimental

2.2.1. Preparation of Resveratrol Solution

A stock solution of *Trans*-resveratrol (Res, M.W. 228.25) was prepared by dissolving resveratrol powder (16 mg) in 1000 µL of dimethyl sulfoxide (DMSO), stirring for 3 min to obtain a concentration of 70 mM. From this solution, 143 µL were withdrawn and diluted in 857 µL of bidistilled sterile water under stirring for 3 min to obtain a concentration of 10 mM. Both solutions were stored in dark glass vials at −20 °C until use, and 4 or 8 µL of the second solution was added to the *cis*-side of the membrane, to obtain the final concentration of 10 or 20 µM, respectively.

2.2.2. Preparation of PLMs

Channel-like activities were recorded in PLMs of POPC or POPC:Ch (65:35, *w/w*) or DOPS:DOPE (1:2, *w/w*) in 1% of n-decane. Bilayers were formed across a 300-µm hole in a Teflon partition separating two Teflon chambers (volume 4000 µL) which held symmetrical 0.1 M KCl solutions, pH = 7, temperature 23 ± 1 °C. The aqueous solutions were used unbuffered. The salts used in the experiments were of analytical grade. The Müller-Rudin or painted technique was used to form PLMs with lipids solubilized in n-decane [45,46]. Briefly, a small volume of 0.5–1 µL of lipid solution is applied through a micropipette directly onto the hole of the Teflon set; a PLM forms in ~10 min after draining the excess solvent into the aqueous bathing solution. In all experiments performed, the conductance and capacitance of each membrane was tested by applying a voltage of ±120 mV for 15–20 min under stirring to ensure that the membrane was stable.

2.2.3. Determination of Conductance of Channel-Like Events

To determine conductance, we measured the amplitude of channel-like events by hand. The single channel data, filtered at 300 Hz, were obtained from at least three experiments (more than 100 single channel-like events) performed on different days. A histogram of the conductance amplitude distribution for each experiment was constructed and fitted by a Gaussian distribution function. Results are expressed as central conductance ±standard error ($\Lambda_c \pm$ SE) and were evaluated by analysis of variance (ANOVA-Tukey test) and Student t test. A value of $p < 0.05$ was considered significant. The Gaussian distribution function, ANOVA test, Student t test, and the fitting procedures were performed using the GraphPad Prism 3 software (GraphPad Prism^{TM} version 3.0, GraphPad Software, San Diego, CA, USA).

To determine the frequency (number of channel-like events in 60 s), any detection of channel events was counted as successful. Results are expressed as frequency ± standard deviation (F ± SD).

2.2.4. Determination of PLM Capacitance

The method used to determine PLM capacitance was as reported in a previous study [47]. Briefly, membrane capacitance was calculated using a calibration curve obtained by simulating the membrane capacitance with a discrete set of capacitances of known values, C_n, and measuring the corresponding output voltage, V_{lh}. The data obtained were fitted by the formula:

$$V_{lh} = \frac{A \times C_n}{(B + C_n)},$$

in which A and B are free parameters to be estimated by the fitting procedures. The values of parameters A and B were used to transform the V_{lh} value into capacitance data.

3. Results

3.1. Membrane Stability

Many studies have shown that resveratrol interacts with membrane lipids due to its lipophilicity. Resveratrol's localization in membranes or membrane models is a matter of debate.

In this study, we used a classic electrophysiological method to investigate resveratrol's microscopic incorporation at concentrations of 10 and 20 µM, in PLMs of different composition. Channel-like event formation indicates that molecules of a substance are incorporating into the membrane. PLMs of different composition were used to see whether resveratrol is able to incorporate into PLMs and to form conductive units, and under what experimental conditions. With this aim, POPC was used as a zwitterionic membrane, POPC:Ch as a neutral mixed membrane and DOPS:DOPE as a mixed membrane with diluted charges.

First of all, we tested PLM stability by applying a voltage of ±120 mV for 10–15 min under stirring and by monitoring the constant values for PLM conductance and capacitance. Preliminarily, to exclude any non-specific and destabilizing effects of DMSO per se on the PLMs used, we performed experiments by adding 4 or 8 µL of a H_2O:DMSO (6:1) mixture to the medium, on the *cis* side of the membrane. In many different experiments and independently of lipid composition, the PLMs were stable even when applying a voltage of ±100 mV for 12 h under stirring and monitoring the constant values for PLM conductance and capacitance.

3.2. Resveratrol Interaction with POPC PLMs

In many different experiments on POPC PLMs, the addition of 10 or 20 µM of resveratrol to the *cis* side of the medium facing the membrane was performed at the applied voltage of 100 mV (addition voltage). After different lag times depending on the resveratrol concentration, channel-like activity appears as non-random discrete events that fluctuate between conductive or non-conductive states compatible with channel-type openings and closures with different conductance levels and frequencies. After the first channel-like event, alternating periods of channel-like activity can be observed, during which the number of events can be rigorously analyzed, followed by quiescent periods and periods of paroxystic channel-like activity that often lead to membrane destabilization until rupture. Figure 4 shows a typical example of channel-like event recordings with associated histograms of the conductance fluctuations.

In experiments at the 10 µM resveratrol concentration, channel-like activity appears after about 60 minutes' lag time at an applied voltage of 100 mV. After the first channel-like event, the applied voltage can be lowered as far as 60 mV and channel amplitude can be monitored. The channel-like activity was registered at applied voltages of ±60, ±80, ±100, and ±120 mV, each applied for 1 h, starting from 60 mV.

By doubling the resveratrol concentration to 20 µM, the lag time reduced to 20 min, after which channel-like activity appeared at an applied voltage of 100 mV, manifesting in a paroxystic manner. Paroxystic activity occurred even when the applied voltage was lowered to 80 mV; therefore, the applied voltage was lowered to 60 mV and channel-like activity appeared as discrete current fluctuations with different conductance levels. Channel-like activity was monitored at applied voltages of ±40 and ±60 mV, each applied for 1 h, starting from 60 mV.

Table 1 summarizes the central conductance values (Λ_c ± SE), obtained by fitting the experimental data with a Gaussian function (Figure 4), and frequency values (F ± SD) at the two different concentrations of resveratrol. Interestingly, at the 10 µM resveratrol concentration, the frequency values were higher at negative applied voltages than at positive ones, indicating a higher turnover of channel-like events.

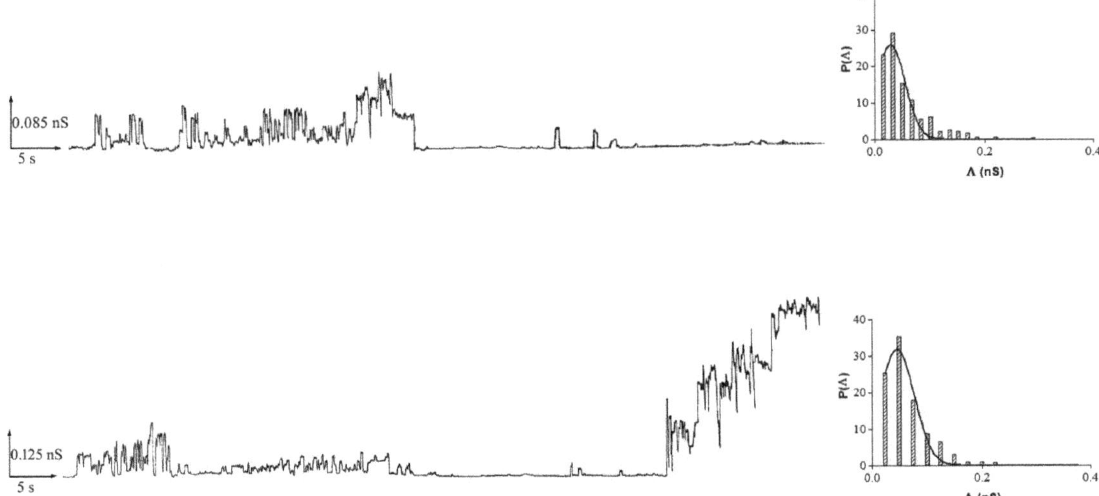

Figure 4. Resveratrol channel-like activity in POPC planar lipid membranes (PLMs). Representative traces illustrating channel activity of resveratrol in membranes made up of POPC with associated histograms of the conductance fluctuations. The histograms of the probability, $P(\Lambda)$, for the frequency of a given conductivity unit were fitted by a Gaussian which is shown as a solid curve. Experiments were performed in the presence of 10 µM (top trace) and 20 µM (bottom trace) of resveratrol added to the *cis* side, while the aqueous phase contained 0.1 M KCl (pH 7) and T = 23 ± 1 °C. Applied voltage was set to 60 mV (top trace) and 40 mV (bottom trace).

Table 1. Characteristic parameters of resveratrol channel-like event in POPC PLM. The mean conductance ($\Lambda c \pm SE$) and frequency ($F \pm SD$) of resveratrol channel-like events at different applied voltages. The minimum and maximum number of channel-like events considered (N) out of a total number of channel-like events considered (Nt) was: 130 < N < 375, Nt = 1640, at the resveratrol concentration of 10 µM. The minimum and maximum number of channel-like events considered (N) out of a total number of channel-like events considered (Nt) was: 280 < N < 622, Nt = 1757, at the resveratrol concentration of 20 µM.

Vs mV	[Res] = 10 µM $\Lambda c \pm SE$ nS	[Res] = 10 µM $F \pm SD$	[Res] = 20 µM $\Lambda c \pm SE$ nS	[Res] = 20 µM $F \pm SD$
120	0.012 ± 0.003	12.28 ± 1.08		
100	0.022 ± 0.002	13.90 ± 1.48		
80	0.019 ± 0.006	14.47 ± 0.74		
60	0.029 ± 0.004	9.88 ± 0.52	0.034 ± 0.001	12.20 ± 0.55
40			0.046 ± 0.003	13.03 ± 0.52
−40			0.054 ± 0.003	12.50 ± 1.01
−60	0.030 ± 0.003	16.97 ± 0.98	0.034 ± 0.001	9.71 ± 0.51
−80	0.016 ± 0.008	30.28 ± 4.16		
−100	0.020 ± 0.001	21.93 ± 1.35		
−120	0.016 ± 0.0006	18.78 ± 2.21		

Another parameter that we calculated is the duration of observed channel-like events. The duration of channel-like events was within a range from 1.25 to 2.25 s and from 0.75 to 1.25 s when the resveratrol concentration was 10 and 20 µM, respectively. The lifetime for two resveratrol concentrations, although not significantly different, indicates that channel-like events were more stable at lower resveratrol concentrations than at higher ones.

The results obtained with POPC PLMs indicate that resveratrol is able to penetrate into the hydrophobic core of PLMs and to form channel-like events. The lag time, the applied voltages at which the channel-like activity manifests, and the shorter lifetimes decrease on increasing resveratrol concentration, in agreement with its poor water solubility and its partition coefficient and with the concept that an appropriate resveratrol/lipid ratio, is required to permeabilize the membrane.

Our experimental apparatus can measure conductance and capacitance simultaneously. Capacitance is related to the property of a membrane to act as a capacitor and is inversely proportional to its thickness. The capacitance parameter is considered to be the best tool for probing the stability and formal goodness of the double lipid layer before and after any experiment concerning the incorporation of a different substance can be conducted [47].

In all experiments, before the addition of resveratrol, the basic capacitance of the POPC PLMs was in the range of 0.27–0.28 $\mu F/cm^2$, remaining constant. After the addition of resveratrol, we observed that the capacitance decreased during the lag time, then increased when the first channel-like events appeared, reaching higher values than those of basic capacitance; these values remain almost constant all the way through to the end of the experiment. This behavior was independent of the resveratrol concentrations used in this study.

Table 2 reports the mean capacitance (C $\pm$ SE) values calculated at different times before and after the addition of resveratrol. To standardize the results, capacitance was measured just before the addition of resveratrol (T0), 70% of the way through the lag time (T1), at the end of the lag time when the channel-like activity appears (T2), and after an average of 1 h from the first channel-like event when the capacitance signal was almost constant (T3). The capacitance variation observed could be due, initially, to resveratrol adsorption to the planar lipid bilayer and, subsequently, to its insertion into the bilayer once an appropriate resveratrol/lipid ratio had been reached.

Table 2. Capacitance variation in POPC PLM. Mean values of the membrane capacitance (C $\pm$ SE) calculated at T0 (just before the addition of resveratrol), T1 (at the 70% of lag time), T2 (at the end of lag time), and T3 (after an average of 1 h from first channel-like event appearance). The mean value was obtained from at least three experiments.

Time	C $\pm$ SE $\mu F/cm^2$ [Res] = 10 μM	C $\pm$ SE $\mu F/cm^2$ [Res] = 20 μM
T0	0.27 $\pm$ 0.01	0.28 $\pm$ 0.01
T1	0.11 $\pm$ 0.002	0.10 $\pm$ 0.003
T2	0.28 $\pm$ 0.02	0.29 $\pm$ 0.01
T3	0.30 $\pm$ 0.02	0.31 $\pm$ 0.02

3.3. Resveratrol Interaction with POPC:Ch PLM

The addition of cholesterol to the POPC solution used to form membranes results in channel-like activity, indicating that the sterol does not prevent the insertion of resveratrol molecules into PLMs made up of POPC:Ch.

The addition voltage of resveratrol was 100 mV both for experiments at a concentration of 10 μM and for those at a concentration of 20 μM. The lag time and applied voltage in which the channel-like activity appears depend on the resveratrol concentration used in the two series of experiments.

After the addition of resveratrol to the *cis* side of the medium facing the membrane, channel-like activity manifests, first of all, as a paroxystic activity then characterized by current jumps compatible with channel-type openings and closures with different conductance levels followed, sometimes, by quiescent periods. Figure 5 shows an example of chart recordings of resveratrol's channel-like activity when incorporated into

POPC:Ch PLMs at the applied voltage of 80 and 40 mV with associated histograms of the conductance fluctuations.

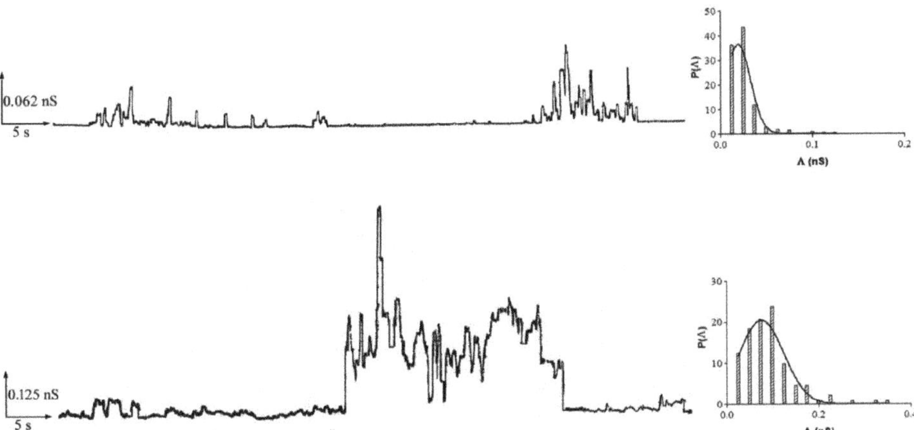

Figure 5. Resveratrol channel-like activity in POPC:Ch PLMs. Representative traces illustrating channel activity of resveratrol in membranes made up of POPC:Ch (65:35, w:w) with associated histograms of the conductance fluctuations. The histograms of the probability, $P(\Lambda)$, for the frequency of a given conductivity unit were fitted by a Gaussian which is shown as a solid curve. Experiments were performed in the presence of 10 µM (top trace) and 20 µM (bottom trace) of resveratrol added to the *cis* side, while the aqueous phase contained 0.1 M KCl (pH 7) and T = 23 ± 1 °C. Applied voltage was set to 80 mV (top trace) and 40 mV (bottom trace).

In the set of experiments at 10 µM of resveratrol, channel-like activity appears after a lag time of 6–7 h at an applied voltage of 120 mV (activation voltage) as a paroxystic activity, after which the applied voltage was lowered down as far as 80 mV where the channel-like event amplitude could be monitored. The channel-like activity was recorded at applied voltages of ±80, ±100, and ±120 mV, each applied for 1 h, starting from 80 mV.

In the set of experiments at 20 µM of resveratrol, channel-like activity manifests as a paroxystic activity after a lag time of 2 h at an applied voltage of 100 mV followed by membrane rupture. After PLM breakage and withdrawal, the channel-like activity occurred spontaneously after a lag time of about 15 min at an applied voltage of 40 mV. The channel-like activity was monitored at applied voltages of ±40 mV, each applied for 90 min, starting from 40 mV. The application of voltages higher than ±40 mV (±60 and ±80 mV) determined paroxystic activity followed by destabilization of the membrane.

Table 3 reports the biophysical and statistic parameters (Λ_c ± SE and F ± SD, respectively) of the channel-like events obtained under the two different experimental conditions. It is important to note that, at a resveratrol concentration of 10 µM, the central conductance and frequency values seem to be independent of applied voltages both at positive and at negative voltages.

The comparatively higher lag times and applied voltages observed in these experimental sets than those for POPC PLMs may be due to the effect of cholesterol on acyl carbon chains that are more closely packed. Resveratrol penetrates the hydrophobic core of the membrane less easily.

As was done for POPC PLMs, we calculated the lifetime of channel-like events formed by resveratrol. The duration of channel-like events was within a range from 1.25 to 1.75 s when the resveratrol concentration was 10 µM, while, at 20 µM, the duration of channel-like events was 1.75 s at applied voltages of ±40 mV.

Table 3. Characteristic parameters of resveratrol channel-like event in POPC:Ch PLM. The mean conductance ($\Lambda c \pm SE$) and frequency ($F \pm SD$) of resveratrol channel-like events at different applied voltages. The minimum and maximum number of channel-like events considered (N) out of a total number of channel-like events considered (Nt) was: $186 < N < 516$, at the resveratrol concentration of 10 µM. The minimum and maximum number of channel-like events considered (N) out of a total number of channel-like events considered (Nt) was: $265 < N < 337$, Nt = 602, at the resveratrol concentration of 20 µM.

Vs mV	[Res] = 10 µM		[Res] = 20 µM	
	$\Lambda c \pm SE$ nS	$F \pm SD$	$\Lambda c \pm SE$ nS	$F \pm SD$
120	0.020 ± 0.001	11.10 ± 0.49		
100	0.020 ± 0.001	9.63 ± 0.62		
80	0.019 ± 0.002	6.80 ± 0.47		
40			0.065 ± 0.003	9.87 ± 0.86
−40			0.050 ± 0.003	6.95 ± 0.70
−80	0.020 ± 0.002	10.78 ± 0.79		
−100	0.021 ± 0.0006	12.74 ± 0.60		
−120	0.016 ± 0.0004	7.20 ± 1.07		

The capacitance behavior recorded in the experiments with POPC:Ch PLMs was similar to that observed for POPC PLMs at the two different resveratrol concentrations. Table 4 reports the mean capacitance values at four different times, T0, T1, T2, and T3. It is interesting to note that the low values of capacitance measured at 70% of lag time (T1) could be due to resveratrol adsorption to the POPC:Ch PLM before its insertion into the bilayer, similarly to that observed for POPC PLMs.

Table 4. Capacitance variation in POPC:Ch PLM. Mean values of the membrane capacitance ($C \pm SE$) calculated at T0, T1, T2, and T3, for which meanings are reported in the legend to Table 3. The mean value was obtained from at least four experiments.

Time	$C \pm SE$ µF/cm^2 [Res] = 10 µM	$C \pm SE$ µF/cm^2 [Res] = 20 µM
T0	0.30 ± 0.02	0.30 ± 0.02
T1	0.11 ± 0.002	0.12 ± 0.002
T2	0.29 ± 0.01	0.28 ± 0.01
T3	0.30 ± 0.02	0.31 ± 0.01

3.4. Resveratrol Interaction with DOPS:DOPE PLMs

In order to test the role of the polar head and the effect of negative membrane charge on resveratrol incorporation and channel-like events, we carried out experiments using PLMs made up of DOPS:DOPE (1:2, w/w) in which the negative charge is diluted. This membrane is characterized by the presence of the net negative charge of DOPS and the neutral DOPE, a non-lamellar-forming lipid.

The resveratrol concentrations tested were 10 and 20 µM. The addition of resveratrol to the *cis* side of the medium facing the membrane was made at an applied voltage of 100 mV, using the same protocol as that described for POPC and POPC:Ch PLMs.

In all experiments with DOPS:DOPE membranes and independently of the resveratrol concentration, no current fluctuations were observed for many hours using a wide range of applied voltages. No channel-like activity occurred after PLM breakage and the formation of the second membrane by painting the lipid solution present around the hole.

In addition, in these experimental sets, we monitored capacitance. In all experiments, before the addition of resveratrol, the basic capacitance of the PLMs remained constant in the range of 0.26–0.28 µF/cm^2. After the addition of resveratrol, we observed that the capacitance decreased during the lag time until reaching a minimum value that remained constant until the end of the experiment. This behavior was observed both at 10 and 20 μM of resveratrol. As in the experiments with POPC and POPC:Ch PLMs, we calculated the mean capacitance values at the different times, T0, T1, T2, and T3, which are reported in Table 5. The capacitance values remain low (0.11 ± 0.03 μF/cm^2) even after PLM breakage and formation of the second membrane.

Table 5. Capacitance variation in DOPS:DOPE PLM. Mean values of the membrane capacitance (C $\pm$ SE) calculated at T0, T1, T2, and T3, for which meanings are reported in the legend to Table 3. The mean value was obtained from at least six experiments.

Time	C $\pm$ SE μF/cm^2 [Res] = 10 μM	C $\pm$ SE μF/cm^2 [Res] = 20 μM
T0	0.26 ± 0.01	0.28 ± 0.01
T1	0.10 ± 0.003	0.11 ± 0.002
T2	0.10 ± 0.002	0.11 ± 0.002
T3	0.10 ± 0.003	0.12 ± 0.001

These results indicate that resveratrol is unable to incorporate into PLMs made up of DOPS:DOPE and to form channel-like events, probably due to the presence of DOPE, a non-lamellar-forming lipid, as resveratrol may induce a negative curvature of the membrane. The combination of these factors prevent resveratrol incorporation into PLMs; however, we cannot exclude an effect of negatively charged lipid DOPS.

4. Discussion

In recent decades, natural derivatives have been the subject of intensive studies regarding their biological and pharmacological properties. Resveratrol is known for its numerous beneficial effects on human health, even though its mechanisms of action are still unclear. Due to its lipophilicity and poor water solubility, the lipid membrane is the principal target of resveratrol.

An important aspect of resveratrol's interaction with the membrane is that it induces changes in the membrane's biophysical properties. The resveratrol molecule may induce changes either to the interfacial structure or to the chain region, thus modifying membrane packing. Therefore, it is very important to be able to clarify the exact location in or on the membrane.

In this work, by simultaneously measuring membrane conductance and capacitance, we show that resveratrol interacts with a planar lipid membrane and the mode of interaction is due to the bilayer's lipid composition. To our knowledge, this is the first study in which resveratrol's membrane activity has been monitored by electrophysiological measurements, which are not so frequently used for the characterization of drug-membrane interactions and by PLMs that complement studies performed with liposomes.

As is well known, conductance is related to the ionic current through the bilayer when a substance (a protein, peptide, or drug) incorporates into the membrane forming conductive units that span the bilayer, while capacitance is related to the property of a membrane to act as a capacitor; indeed, the lipid bilayer is an insulator separating two electrolytic media. Capacitance is directly proportional to the area of the membrane and inversely proportional to its thickness. Capacitance also depends on the characteristics of the lipids used as the insulating material.

Several studies show that the structure of lipids determines the biophysical properties of membrane (e.g., organization in bilayers, fluidity, hexagonal phases, etc.) [48]. In our work, we used POPC that is prone to form a bilayer [49,50], cholesterol that has effects on membrane fluidity, and DOPE that is prone to form inverse hexagonal phases [49,50]. During the interaction of resveratrol with the planar lipid bilayer, the variation in capacitance can help to shed light on its mechanism of action.

Our results show that resveratrol incorporates into PLMs made up of POPC and forms transient channel-like events in which lag time depends on its concentration. The lag time is the time at which a conductance variation first occurs after resveratrol addition. The different duration of lag time may be due to reaching an appropriate resveratrol/lipid ratio. Once a threshold resveratrol/lipid value has been reached, resveratrol forms transient conductive units permeabilizing the membrane. The capacitance variation, induced by adding resveratrol, may be due to: adsorption of resveratrol (capacitance decrease) occurring before formation of conductive units and/or formation of channel-like events (capacitance increase) across the membrane. The paroxystic activity reported above may be due to the collapse of the conductive units inducing rapid flip-flopping of the membrane lipids. It is important to note that both the lag time and the applied voltages at which the paroxystic activity appears depend on the resveratrol concentration, thus strengthening the concept that an appropriate resveratrol/lipid ratio has been reached.

The addition of cholesterol, a known component of the cellular membrane, to PLMs made up of POPC increases the lag time compared to that observed for experiments with POPC PLMs, regardless of the resveratrol concentration used. The capacitance behaves in a similar fashion to that observed for experiments with PLMs in the absence of cholesterol, decreasing immediately after resveratrol addition and increasing when the channel-like activity appears. Therefore, also in POPC:Ch PLMs, the formation of channel-like events comes after the resveratrol adsorption phase. However, the higher lag times than those obtained for POPC PLMs would seem to indicate that cholesterol makes it more difficult for resveratrol to incorporate into PLMs and to form conductive units, probably due to the tighter packing of the phospholipid tails.

According to some studies, the β-hydroxyl group attached to C3 of the cholesterol molecule, is located in the headgroup region of POPC, while the isooctyl chain, attached to C17, is located deep within the hydrophobic core of the bilayer close to the double bonds of fatty acids [24]. Some authors [1,25] have shown that resveratrol adopts the same position in the membrane as cholesterol. Besides, the higher applied voltage (120 mV) needed to obtain channel-like activity, compared with POPC PLMs, seems to confirm this idea.

Our results appear to be in line with those obtained by other authors [25,29,51] who have shown that resveratrol penetrates into the hydrophobic core of the bilayer close to the double bonds of polyunsaturated fatty acids.

This mode of interaction with lipids might explain the anti-peroxidation effects displayed by resveratrol, because it protects the phospholipids from oxidation reactions, acting as an electron donor to a free radical, neutralizing it and inhibiting its capacity to damage the cell membrane [52].

In contrast with its effects in neutral PLMs, resveratrol does not interact with DOPE PLMs to which negatively charged lipid DOPS have been added. To our knowledge, the interaction of resveratrol with the lipid system containing these phospholipids has never previously been studied.

In the experiments with DOPS:DOPE PLMs, the capacitance decreases after resveratrol addition reaching a minimal value that is quite constant until the end of the experiment.

The behavior of the capacitance and the absence of channel-like activity indicate that resveratrol remains on the membrane surface and is unable to incorporate into the PLM. This behavior may be due to the chemical characteristics of the lipids forming the bilayer and/or to the effect of resveratrol on the membrane surface, which promote the formation of the inverse hexagonal phase and of membrane negative curvature.

Several studies have shown that the nature and molecular shape of lipids determine their intrinsic tendency to form distinct phases [49] and to affect membrane curvature. Phosphatidylethanolamine (PE) has a truncated cone shape because its headgroup has a smaller area than the cross-section of its hydrocarbon chains, while phosphatidylcholine (PC) has a cylindrical molecular shape. Because of these features, PE is prone to form an inverse hexagonal phase and PC bilayer. Besides, dioleoyl lipids with phosphatidylethanolamine headgroups adopt a negatively-curved surface because the PE headgroup is smaller than the PC headgroup [53], and they are able to interact most with the glycerol and phosphate groups of neighboring lipids, probably due to the differing hydration properties influencing the curvature [54].

On the other hand, it has been demonstrated that many substances that are active at the membrane interface induce membrane curvature, namely local or global deformations of the membrane.

Numerous studies show that there are many substances, interfacially active peptides [55–57] and non-peptide compounds that can influence membrane curvature, and vice versa. Yesylevskyy and colleagues have shown that the permeability of the model lipid membrane for cisplatin and gemcitabine depends on the curvature [58]. Barry and colleagues showed that curcumin, a lipophilic drug, binds to dimyristoyl-phosphocholine (DMPC) with a transbilayer orientation, anchoring to the bilayer by a hydrogen bond to the phosphate group, while, in dipalmitoleoyl-glycero-phosphoethanolamine (DiPOPE) vesicles, it promotes and stabilizes the formation of the inverted hexagonal (H_{II}) phase. According to the authors, this feature of curcumin is indirect evidence that it induces negative membrane curvature [59].

On the other hand, primaquine, a potent antimalarial agent, increases the phase transition temperature (T_H) of Lα and reduces the transition enthalpy of palmitoyl-oleoyl-phosphatidylethanolamine (POPE) vesicles, indicating that primaquine promotes positive membrane curvature and stabilizes the fluid phase of POPE vesicles [60].

All these studies indicate that many substances, peptides, and drugs induce modifications of a membrane's biophysical properties caused by the chemical characteristics of the substances in question. In line with this concept, our work shows that resveratrol interacts with model membranes, modifying their biophysical parameters for different lipid compositions. Studies on drug-membrane interaction are very important in order to understand and clarify their mechanism of action.

5. Conclusions

Our results indicate that resveratrol interacts with zwitterionic and neutral PLMs, by incorporating into PLMs and forming transient conductive units, whereas it is unable to interact with DOPE:DOPS PLMs. In light of this, we hypothesize a model in which its fundamental steps are: (1) adsorption of resveratrol onto the membrane surface. This step is common to the neutral PLMs and to the negatively-charged PLMs used in this study; and (2) in neutral PLMs, formation of transient conductive units made up of molecules of resveratrol oriented with their long axis along the membrane normal and assembled together (Figure 6). The incorporation and channel-like events formation indicate that resveratrol is located in the hydrophobic core of the bilayer. These results, obtained by simultaneously monitoring membrane conductance and capacitance, may help to clarify the mechanism of action by which resveratrol performs its beneficial effects on human health, such as its antioxidant or anti-microbial properties and its scavenging of lipid radicals.

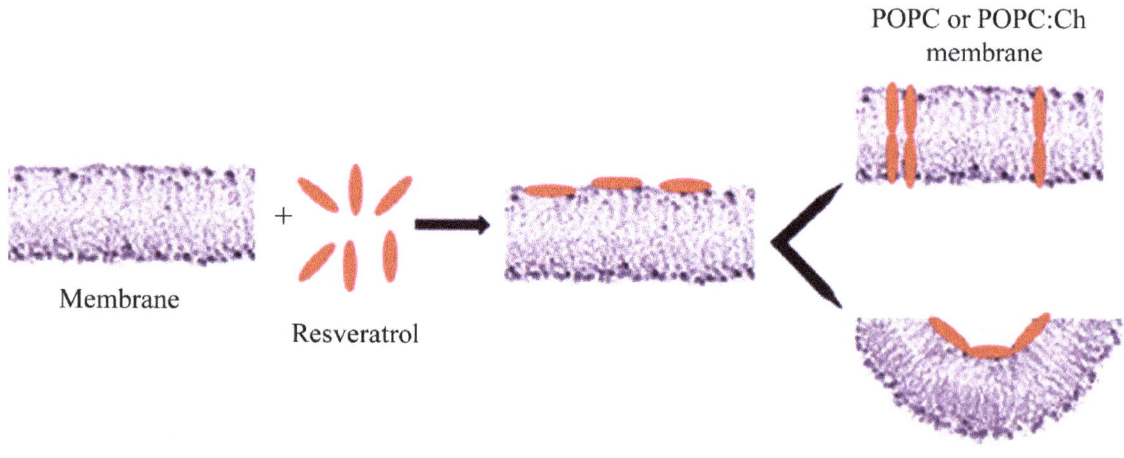

Figure 6. Schematic model of the resveratrol interaction with PLM at different lipid composition. Resveratrol adsorbs onto the membrane surface, regardless to the PLM composition. In POPC or POPC:Ch PLMs (top), resveratrol inserts and assembles into membrane forming conductive units, once an appropriate resveratrol/lipid ratio has been reached. In DOPE:DOPS PLMs (bottom), resveratrol inducing negative curvature of membrane is unable to insert into it.

Funding: This research did not receive any specific grant from funding agencies in the public, commercial, or not-for-profit sectors.

Acknowledgments: The authors acknowledge Anthony Green for proofreading and providing linguistic advice.

Conflicts of Interest: The author declares no conflict of interest.

References

1. Brittes, J.; Lúcio, M.; Nunes, C.; Lima, J.L.; Reis, S. Effects of Resveratrol on Membrane Biophysical Properties: Relevance for Its Pharmacological Effects. *Chem. Phys. Lipids* **2010**, *163*, 747–754. [CrossRef]
2. Vestergaard, M.; Ingmer, H. Antibacterial and Antifungal Properties of Resveratrol. *Int. J. Antimicrob. Agents* **2019**, *53*, 716–723. [CrossRef] [PubMed]
3. Bo, S.; Ciccone, G.; Castiglione, A.; Gambino, R.; De Michieli, F.; Villois, P.; Durazzo, M.; Perin, C.P.; Cassader, M. Anti-inflammatory and Antioxidant Effects of Resveratrol in Healthy Smokers a Randomized, Double-Blind, Placebo-Controlled, Cross-over Trial. *Curr. Med. Chem.* **2013**, *20*, 1323–1331. [CrossRef] [PubMed]
4. Lançon, A.; Frazzi, R.; Latruffe, N. Anti-Oxidant, Anti-Inflammatory and Anti-Angiogenic Properties of Resveratrol in Ocular Diseases. *Molecules* **2016**, *21*, 304. [CrossRef]
5. Athar, M.; Back, J.H.; Kopelovich, L.; Bickers, D.R.; Kim, A.L. Multiple Molecular Targets of Resveratrol: Anti-Carcinogenic Mechanisms. *Arch. Biochem. Biophys.* **2009**, *486*, 95–102. [CrossRef] [PubMed]
6. Ko, J.H.; Sethi, G.; Um, J.Y.; Shanmugam, M.K.; Arfuso, F.; Kumar, A.P.; Bishayee, A.; Ahn, K.S. The Role of Resveratrol in Cancer Therapy. *Int. J. Mol. Sci.* **2017**, *18*, 2589. [CrossRef]
7. Baxter, R.A. Anti-aging Properties of Resveratrol: Review and Report of a Potent New Antioxidant Skin Care Formulation. *J. Cosmet. Dermatol.* **2008**, *7*, 2–7. [CrossRef] [PubMed]
8. Wen, S.; Zhang, J.; Yang, B.; Elias, P.M.; Man, M.Q. Role of Resveratrol in Regulating Cutaneous Functions. *Evid. Based Complement. Alternat. Med.* **2020**, *2020*, 2416837. [CrossRef]
9. Braidy, N.; Jugder, B.E.; Poljak, A.; Jayasena, T.; Mansour, H.; Nabavi, S.M.; Sachdev, P.; Grant, R. Resveratrol as a Potential Therapeutic Candidate for the Treatment and Management of Alzheimer's Disease. *Curr. Top. Med. Chem.* **2016**, *16*, 1951–1960. [CrossRef]
10. Granzotto, A.; Zatta, P. Resveratrol and Alzheimer's Disease: Message in a Bottle on Red Wine and Cognition. *Front. Aging Neurosci.* **2014**, *6*, 95. [CrossRef]

11. Hartman, R.E.; Shah, A.; Fagan, A.M.; Schwetye, K.E.; Parsadanian, M.; Schulman, R.N.; Finn, M.B.; Holtzman, D.M. Pomegranate Juice Decreases Amyloid Load and Improves Behavior in a Mouse Model of Alzheimer's Disease. *Neurobiol. Dis.* **2006**, *24*, 506–515. [CrossRef]

12. Koukoulitsa, C.; Villalonga-Barber, C.; Csonka, R.; Alexi, X.; Leonis, G.; Dellis, D.; Hamelink, E.; Belda, O.; Steele, B.R.; Micha-Screttas, M.; et al. Biological and Computational Evaluation of Resveratrol Inhibitors against Alzheimer's Disease. *J. Enzyme Inhib. Med. Chem.* **2016**, *31*, 67–77. [CrossRef]

13. Ono, K.; Yoshiike, Y.; Takashima, A.; Hasegawa, K.; Naiki, H.; Yamada, M. Potent Anti-Amyloidogenic and Fibril-Destabilizing Effects of Polyphenols in Vitro: Implications for the Prevention and Thera-Peutics of Alzheimer's Disease. *J. Neurochem.* **2003**, *87*, 172–181. [CrossRef] [PubMed]

14. Pasinetti, G.M.; Wang, J.; Ho, L.; Zhao, W.; Dubner, L. Roles of Resveratrol and Other Grape-Derived Polyphenols in Alzheimer's Disease Prevention and Treatment. *Biochim. Biophys. Acta* **2015**, *1852*, 1202–1208. [CrossRef] [PubMed]

15. Rege, S.D.; Geetha, T.; Griffin, G.D.; Broderick, T.L.; Babu, J.R. Neuroprotective Effects of Resveratrol in Alzheimer Disease Pathology. *Front. Aging Neurosci.* **2014**, *6*, 218. [CrossRef]

16. Bastianetto, S.; Ménard, C.; Quirion, R. Neuroprotective Action of Resveratrol. *Biochim. Biophys. Acta* **2015**, *1852*, 1195–1201. [CrossRef] [PubMed]

17. Yashiro, T.; Nanmoku, M.; Shimizu, M.; Inoue, J.; Sato, R. Resveratrol Increases the Expression and Activity of the Low Density Lipoprotein Receptor in Hepatocytes by the Proteolytic Activation of the Sterol Regulatory Element-Binding Proteins. *Atherosclerosis* **2012**, *220*, 369–374. [CrossRef]

18. Voloshyna, I.; Hussaini, S.M.; Reiss, A.B. Resveratrol in Cholesterol Metabolism and Atherosclerosis. *J. Med. Food* **2012**, *15*, 763–773. [CrossRef]

19. Galfi, P.; Jakus, J.; Molnar, T.; Neogrady, S.; Csordas, A. Divergent Effects of Resveratrol, a Polyphenolic Phytostilbene, on Free Radical Levels and Type of Cell Death Induced by the Histone Deacetylase Inhibitors Butyrate and Trichostatin A. *J. Steroid Biochem. Mol. Biol.* **2005**, *94*, 39–47. [CrossRef]

20. Frankel, E.N.; Waterhouse, A.L.; Kinsella, J.E. Inhibition of Human LDL Oxidation by Resveratrol. *Lancet* **1993**, *341*, 1103–1104. [CrossRef]

21. Jensen, M.D.; Sheng, W.; Simonyi, A.; Johnson, G.S.; Sun, A.Y.; Sun, G.Y. Involvement of Oxidative Pathways in Cytokine-Induced Secretory Phospholipase A2-IIA in Astrocytes. *Neurochem. Int.* **2009**, *55*, 362–368. [CrossRef]

22. Sun, G.Y.; Shelat, P.B.; Jensen, M.B.; He, Y.; Sun, A.Y.; Simonyi, A. Phospholipases a2 and Inflammatory Responses in the Central Nervous System. *Neuromol. Med.* **2010**, *12*, 133–148. [CrossRef]

23. Yarla, N.S.; Bishayee, A.; Sethi, G.; Reddanna, P.; Kalle, A.M.; Dhananjaya, B.L.; Dowluru, K.S.; Chintala, R.; Duddukuri, G.R. Targeting Arachidonic Acid Pathway by Natural Products for Cancer Prevention and Therapy. *Semin. Cancer Biol.* **2016**, *40*, 48–81. [CrossRef]

24. Fei, Q.; Kent, D.; Botello-Smith, W.M.; Nur, F.; Nur, S.; Alsamarah, A.; Chatterjee, P.; Lambros, M.; Luo, Y. Molecular Mechanism of Resveratrol's Lipid Membrane Protection. *Sci. Rep.* **2018**, *8*, 1587. [CrossRef] [PubMed]

25. Fabris, S.; Momo, F.; Ravagnan, G.; Stevanato, R. Antioxidant Properties of Resveratrol and Piceid on Lipid Peroxidation in Micelles and Monolamellar Liposomes. *Biophys. Chem.* **2008**, *135*, 76–83. [CrossRef]

26. Ghellinck, d.A.; Shen, C.; Fragneto, G.; Klösgen, B. Probing the Position of Resveratrol in Lipid Bilayers: A Neutron Reflectivity Study. *Colloids Surf. B Biointerfaces* **2015**, *134*, 65–72. [CrossRef] [PubMed]

27. Han, J.; Suga, K.; Hayashi, K.; Okamoto, Y.; Umakoshi, H. Multi-Level Characterization of the Membrane Properties of Resveratrol-Incorporated Liposomes. *J. Phys. Chem. B* **2017**, *121*, 4091–4098. [CrossRef]

28. Cardia, M.C.; Caddeo, C.; Lai, F.; Fadda, A.M.; Sinico, C.; Luhmer, M. H NMR Study of the Interaction of Trans-Resveratrol with Soybean Phosphatidylcholine Liposomes. *Sci. Rep.* **2019**, *9*, 17736. [CrossRef]

29. Bonechi, C.; Martini, S.; Ciani, L.; Lamponi, S.; Rebmann, H.; Rossi, C.; Ristori, S. Using Liposomes as Carriers for Polyphenolic Compounds: The Case of Trans-Resveratrol. *PLoS ONE* **2012**, *7*, e41438. [CrossRef] [PubMed]

30. Balanc, B.; Ota, A.; Djordjevic, V.; Sentjurc, M.; Nedovic, V.; Bugarski, B.; Poklar, U.N. Resveratrol-Loaded Liposomes: Interaction of Resveratrol with Phospholipids. *Eur. J. Lipid Sci. Technol.* **2015**, *117*, 1615–1626. [CrossRef]

31. Neves, A.R.; Nunes, C.; Reis, S. Resveratrol Induces Ordered Domains Formation in Biomembranes: Implication for Its Pleiotropic Action. *Biochim. Biophys. Acta* **2016**, *1858*, 12–18. [CrossRef]

32. Neves, A.R.; Nunes, C.; Reis, S. New Insights on the Biophysical Interaction of Resveratrol with Biomembrane Models: Relevance for Its Biological Effects. *J. Phys. Chem. B* **2015**, *119*, 11664–11672. [CrossRef]

33. Martinović, N.; Abramovič, H.; Poklar, U.N. Inhibition of Copper-Induced Lipid Peroxidation by Sinapic Acid and Its Derivatives in Correlation to Their Effect on the Membrane Structural Properties. *Biochim. Biophys. Acta Biomembr.* **2019**, *1861*, 1–8. [CrossRef] [PubMed]

34. Andrade, S.; Ramalho, M.; Loureiro, A.; Pereira, M. The Biophysical Interaction of Ferulic Acid with Liposomes as Biological Membrane Model: The Effect of the Lipid Bilayer Composition. *J. Mol. Liq.* **2020**. [CrossRef]

35. Gutiérrez, M.E.; García, A.F.; Madariaga, A.M.; Sagrista, M.L.; Casadó, F.J.; Mora, M. Interaction of Tocopherols and Phenolic Compounds with Membrane Lipid Components: Evaluation of Their Antioxidant Activity in a Liposomal Model System. *Life Sci.* **2003**, *72*, 2337–2360. [CrossRef]

36. Arora, A.; Byrem, T.M.; Nair, M.G.; Strasburg, G.M. Modulation of Liposomal Membrane Fluidity by Flavonoids and Isoflavonoids. *Arch. Biochem. Biophys.* **2000**, *373*, 102–109. [CrossRef] [PubMed]
37. Tsuchiya, H.; Nagayama, M.; Tanaka, T.; Furusawa, M.; Kashimata, M.; Takeuchi, H. Membrane-Rigidifying Effects of Anti-cancer Dietary Factors. *Biofactors* **2002**, *16*, 45–56. [CrossRef]
38. Selvaraj, S.; Mohan, A.; Narayanan, S.; Sethuraman, S.; Krishnan, U.M. Dose-Dependent Interaction of Trans-Resveratrol with Biomembranes: Effects on Antioxidant Property. *J. Med. Chem.* **2013**, *56*, 970–981. [CrossRef]
39. Shahane, G.; Ding, W.; Palaiokostas, M.; Orsi, M. Physical Properties of Model Biological Lipid Bilayers: Insights from All-Atom Molecular Dynamics Simulations. *J. Mol. Model.* **2019**, *25*, 76. [CrossRef] [PubMed]
40. Murugova, T.N.; Balgavý, P. Molecular Volumes of DOPC and DOPS in Mixed Bilayers of Multilamellar Vesicles. *Phys. Chem. Chem. Phys.* **2014**, *16*, 18211–18216. [CrossRef] [PubMed]
41. Rand, R.P.; Fuller, N.L. Structural Dimensions and Their Changes in a Reentrant Hexagonal-Lamellar Transition of Phospholipids. *Biophys. J.* **1994**, *66*, 2127–2138. [CrossRef]
42. Suetsugu, S.; Kurisu, S.; Takenawa, T. Dynamic Shaping of Cellular Membranes by Phospholipids and Membrane-Deforming Proteins. *Physiol. Rev.* **2014**, *94*, 1219–1248. [CrossRef] [PubMed]
43. Tien, H.; Ottova, A.-L. *Membrane Biophysics: As Viewed from Experimental Bilayer Lipid Membranes*; Elsevier: Amsterdam, The Netherlands, 2000.
44. Tien, H.; Ottova, A.-L. The Lipid Bilayer Concept and Its Experimental Realization: From Soap Bubbles, Kitchen Sink, to Bilayer Lipid Membranes. *J. Membr. Sci.* **2001**, *189*, 83–117. [CrossRef]
45. Müller, P.; Rudin, D.; Tien, T.; Weacott, W. Reconstitution of Cell Membrane Structure in Vitro and Its Trasformation into an Excitable System. *Nature* **1962**, *194*, 979–980. [CrossRef] [PubMed]
46. Tien, T.; Mountz, J.; Martinosi, A. Protein-Lipid Interaction in Bilayer Lipid Membranes (BLM). In *Enzyme of Biological Membranes*; NY Plenum: New York, NY, USA, 1977; pp. 139–170.
47. Micelli, S.; Gallucci, E.; Meleleo, D.; Stipani, V.; Picciarelli, V. Mitochondrial Porin Incorporation into Black Lipid Membranes: Ionic and Gating Contribution to the Total Current. *Bioelectrochemistry* **2002**, *57*, 97–106. [CrossRef]
48. Faroux, J.M.; Ureta, M.M.; Tymczyszyn, E.E.; Gómez-Zavaglia, A. An Overview of Peroxidation Reactions Using Liposomes as Model Systems and Analytical Methods as Monitoring Tools. *Colloids Surf. B Biointerfaces* **2020**, *195*, 111254. [CrossRef]
49. Koller, D.; Lohner, K. The Role of Spontaneous Lipid Curvature in the Interaction of Interfacially Active Peptides with Membranes. *Biochim. Biophys. Acta* **2014**, *1838*, 2250–2259. [CrossRef] [PubMed]
50. Cullis, P.R.; de Kruijff, B. Lipid Polymorphism and the Functional Roles of Lipids in Biological Membranes. *Biochim. Biophys. Acta* **1979**, *559*, 399–420. [CrossRef]
51. Koukoulitsa, C.; Durdagi, S.; Siapi, E.; Villalonga-Barber, C.; Alexi, X.; Steele, B.R.; Micha, M.-S.; Alexis, M.N.; Tsantili-Kakoulidou, A.; Mavromoustakos, T. Comparison of Thermal Effects of Stilbenoid Analogs in Lipid Bilayers Using Differential Scanning Calorimetry and Molecular Dynamics: Correlation of Thermal Effects and Topographical Position with Antioxidant Activity. *Eur. Biophys. J.* **2011**, *40*, 865–875. [CrossRef]
52. Zhang, J.; Fu, Y.; Yang, P.; Liu, X.; Li, Y.; Gu, Z. ROS Scavenging Biopolymers for Anti-Inflammatory Diseases: Classification and Formulation. *Adv. Mater. Interfaces* **2020**, *7*. [CrossRef]
53. Rand, R.P.; Fuller, N.L.; Gruner, S.M.; Parsegian, V.A. Membrane Curvature, Lipid Segregation, and Structural Transitions for Phospholipids under Dual-Solvent Stress. *Biochemistry* **1990**, *29*, 76–87. [CrossRef] [PubMed]
54. Sodt, A.J.; Pastor, R.W. Molecular Modeling of Lipid Membrane Curvature Induction by a Peptide: More Than Simply Shape. *Biophys. J.* **2014**, *106*, 1958–1969. [CrossRef] [PubMed]
55. Haney, E.F.; Nathoo, S.; Vogel, H.J.; Prenner, E.J. Induction of Non-lamellar Lipid Phases by Antimicrobial Peptides: A Potential Link to Mode of Action. *Chem. Phys. Lipids* **2010**, *163*, 82–93. [CrossRef] [PubMed]
56. Gallucci, E.; Meleleo, D.; Micelli, S.; Picciarelli, V. Magainin 2 Channel Formation in Planar Lipid Membranes: The Role of Lipid Polar Groups and Ergosterol. *Eur. Biophys. J.* **2003**, *32*, 22–32. [CrossRef] [PubMed]
57. Matsuzaki, K.; Mitani, Y.; Akada, K.Y.; Murase, O.; Yoneyama, S.; Zasloff, M.; Miyajima, K. Mechanism of Synergism between Antimicrobial Peptides Magainin 2 and PGLa. *Biochemistry* **1998**, *37*, 15144–15153. [CrossRef]
58. Yesylevskyy, S.; Rivel, T.; Ramseyer, C. Curvature Increases Permeability of the Plasma Membrane for Ions, Water and the Anti-cancer Drugs Cisplatin and Gemcitabine. *Sci. Rep.* **2019**, *9*, 17214. [CrossRef]
59. Barry, J.; Fritz, M.; Brender, J.R.; Smith, P.E.; Lee, D.K.; Ramamoorthy, A. Determining the Effects of Lipophilic Drugs on Membrane Structure by Solid-State NMR Spectroscopy: The Case of the Antioxidant Curcumin. *J. Am. Chem. Soc.* **2009**, *131*, 4490–4498. [CrossRef]
60. Basso, L.G.; Rodrigues, R.Z.; Naal, R.M.; Costa-Filho, A.J. Effects of the Antimalarial Drug Primaquine on the Dynamic Structure of Lipid Model Membranes. *Biochim. Biophys. Acta* **2011**, *1808*, 55–64. [CrossRef]

Article

Experimental and Theoretical Approaches to Describing Interactions in Natural Cell Membranes Occurring as a Result of Fatal Alcohol Poisoning

Aneta D. Petelska [1,*], Michał Szeremeta [2], Joanna Kotyńska [1] and Anna Niemcunowicz-Janica [2]

[1] Faculty of Chemistry, University of Bialystok, Ciolkowskiego 1K, 15-245 Bialystok, Poland; joannak@uwb.edu.pl

[2] Department of Forensic Medicine, Medical University of Bialystok, Waszyngtona St. 13, 15-230 Bialystok, Poland; michalszeremeta@gmail.com (M.S.); anna.janica@umb.edu.pl (A.N.-J.)

[*] Correspondence: aneta@uwb.edu.pl

Abstract: We propose herein a theoretical model describing the effect of fatal ethanol poisoning on the equilibria between cell membranes and the surrounding ions. Using this model, we determined the parameters characterizing the interaction between the electrolyte solution's ions and the functional groups on the blood cells' surface. Via the application of mathematical equations, we calculated the total surface concentrations of the acidic and basic groups, c_A and c_B, and their association constants with solution ions, K_{AH} and K_{BOH}. Using the determined parameters and mathematical equations' values, we calculated the theoretical surface charge density values. We verified the proposed model by comparing these values with experimental data, which were selected based on measurements of the electrophoretic mobility of erythrocyte and thrombocyte membranes. Compatibility of the experimental and theoretical surface charge density values was observed in the range of pH 2–8, while deviations were observed at higher pH values.

Keywords: fatal ethyl alcohol poisoning; surface charge density; microelectrophoresis; acid–base equilibria; erythrocytes; thrombocytes

Citation: Petelska, A.D.; Szeremeta, M.; Kotyńska, J.; Niemcunowicz-Janica, A. Experimental and Theoretical Approaches to Describing Interactions in Natural Cell Membranes Occurring as a Result of Fatal Alcohol Poisoning. *Membranes* **2021**, *11*, 189. https://doi.org/10.3390/membranes11030189

Academic Editor: Francisco Monroy

Received: 28 January 2021
Accepted: 5 March 2021
Published: 9 March 2021

Publisher's Note: MDPI stays neutral with regard to jurisdictional claims in published maps and institutional affiliations.

1. Introduction

Alcoholism is a severe health problem worldwide. Even if it is not itself the direct cause of death, ethanol abuse is associated with many health risks, including poisoning and the greater spread of HIV and tuberculosis. Alcohol negatively affects various types of blood cells and their functions. Ethyl alcohol abusers most often have defective red blood cells; they are destroyed prematurely, leading to anemia. Drinking too much ethyl alcohol can increase stroke risk as it affects the blood coagulation system [1]. Ethyl alcohol is toxic to the bone marrow (it contains blood cell precursors) and all cells circulating in the blood [2]. Microscopic examination of cultured blood cell precursors has shown that exposure of cells to a wide range of ethyl alcohol concentrations causes damage to the membrane surrounding each cell.

Alcoholism can also modify blood clotting or coagulation. Alcohol consumption can interfere with these processes at several levels, causing thrombocytopenia (abnormally low blood platelet count), thrombocytopathy (impaired platelet function), and reduced fibrinolysis. Researchers have suggested that alcohol intoxication itself causes a decrease in blood platelets. Literature data also indicate that alcohol shortens existing thrombocytes' lifespan and may affect the late stage of platelet production. Alcohol abuse also affects the properties of thrombocytes. These platelet abnormalities include impaired platelet aggregation and decreased secretion or activity of platelet-derived proteins involved in blood coagulation [3–7].

Even small changes in the composition of cell membranes can strongly affect their functioning and physicochemical properties. These changes may occur due to the modification of cell membranes by short-chain alcohols like ethanol, a particularly important representative of molecules that can modulate membranes' properties. As a first approximation to understanding the interaction of ethanol with cell membranes, lipid bilayers are considered. This interaction is of biophysical interest as ethyl alcohol can induce the formation of interdigitated bilayer structures and modulate lipids' phase stability. Numerous experimental and computational studies have been performed to determine how ethyl alcohol affects lipid bilayers [8–10]. Although the alcohol has an amphiphilic nature, its hydrophobicity is limited; it can pass through the bilayer. Besides this, ethanol molecules condense near the interface region between lipids and water. –OH groups are positioned in the bilayer interfacial region (forming hydrogen bonds with hydrophilic lipid headgroups), and the hydrocarbon chains face the hydrophobic core of the bilayer. The presence of ethanol in the lipid membrane has a disordering effect on the hydrocarbon chains, giving rise to an increase in the membrane's area per lipid and fluidity.

The interaction of ethyl alcohol with lipid membranes, both natural and model, has been examined for years. Alcohol-induced modifications in erythrocyte membranes have been extensively investigated using different experimental methods such as Electron paramagnetic resonance spectroscopy (EPR) spectroscopy [11,12], fluorescence anisotropy [13,14], gas chromatography [15], and microelectrophoresis [16–18]. Nevertheless, there is limited data concerning the influence of ethanol poisoning on thrombocyte membranes. In the literature, one can find microelectrohoretic studies of cell membranes' electric properties when intoxicated with ethanol, but it should be emphasized that these reports concern rats. The authors found that alcohol significantly enhanced changes in the surface charge density of erythrocyte membranes [17] and liver cell membranes [18]. The obtained results indicate that changes in cell membrane charge connect with changes in membrane composition.

In recent years, studies of the electrical properties of cell membranes have focused mainly on model membranes—monolayers [19], bilayers [20], and supported lipid membranes [21]—while there are few papers concerning natural membranes [22,23]. The results included in this paper are a continuation of the systematic research on the electrical properties of blood cell membranes conducted by Petelska et al. [24–28]. In this work, we analyze the influence of fatal ethanol poisoning on human blood cell membranes' surface charge, which is an important parameter describing equilibria in both model and natural membranes.

Erythrocytes and thrombocytes have relatively simple structures. Thus, they are ideal cellular models to study the alterations of membranes' physicochemical properties under the influence of small amphiphilic solutes, like ethanol. In cellular systems, the toxicity of ethyl alcohol has been suggested to be due to its interactions with membranes [29]. Due to the lack of literature data on the effect of fatal poisoning with ethyl alcohol on the equilibria between the membranes of erythrocytes and thrombocytes and the surrounding environment, we proposed a mathematical model to describe these equilibria. The obtained data indicate the influence of alcohol intoxication on the blood cell membranes. In our opinion, our quantitative description of the properties of erythrocyte and thrombocyte membranes may help to better understand the impact of fatal ethyl alcohol poisoning on the surface properties of blood cell membranes.

2. Materials and Methods

2.1. Materials

The Ethics Review Board of the Medical University of Bialystok approved the conducted research (No. R-I-002/533/2010). During autopsies made at the Department of Forensic Medicine, Medical University of Bialystok, blood (pH ~ 6.9) was obtained from individuals who died due to fatal alcohol poisoning. The conducted experiment was based on ten selective fatal alcohol poisonings (5 men and 5 women; mean age 45.8 years, range 19–65). Blood was taken for tests from the femoral vein and placed in sterile containers, then sent to the Faculty of Chemistry, University of Bialystok for further research. The obtained results were subjected to a comparative analysis with a control group obtained from live individuals at the Blood Donation Center in Białystok.

RBC isolation: The RBCs were isolated from 2 mL of whole blood by centrifugation for 8 min at room temperature at $900\times$ *g*. The thrombocyte-rich plasma supernatant was removed, and the obtained erythrocytes were washed three times with 0.9% NaCl, then centrifuged for 15 min at $3000\times$ *g*. After the final centrifugation, erythrocytes were resuspended in 0.9% NaCl, followed by microelectrophoretic measurement.

Platelet isolation: The thrombocyte-rich plasma was centrifuged for 8 min at $4000\times$ *g*. The plasma supernatant was removed and discarded. Thrombocytes were washed three times with physiological saline solution (0.9% NaCl) and then centrifuged for 15 min at $3000\times$ *g*. Then, thrombocytes were resuspended in 0.9% NaCl, and microelectrophoretic measurements were made.

The test solutions were made and all purification procedures were performed using ultrapure water from a Milli-Q11 system (18.2 MΩ cm, Millipore, Burlington, MA, USA).

2.2. Methods

Microelectrophoretic Mobility Measurements

The purpose of the microelectrophoretic experiment was to obtain the dependence of the surface charge density on the pH of the electrolyte solution. Using a Zetasizer Nano ZS (Malvern Instruments, Malvern, UK) apparatus allowed us to obtain data on the microelectrophoretic mobility of the blood cells using the laser Doppler microelectrophoresis (LDE) technique. WTW InoLab 720 laboratory pH meter (WTW, Weinheim, Germany) use for all measurements was performed as a function of pH. Six measurements were made (each covering 100–200 series for a duration of 5 s) for each pH value for each sample.

Equation (1) allowed us to calculate of the surface charge density from electrophoretic mobility measurements [30]:

$$\delta = \frac{\eta \cdot u}{d} \tag{1}$$

where η is the viscosity of the solution, u is the electrophoretic mobility, and d is the diffusion layer thickness.

The thickness of the diffusion layer can be determined from the following formula (Equation (2)) [31]:

$$d = \sqrt{\frac{\varepsilon \cdot \varepsilon_0 \cdot R \cdot T}{2 \cdot F^2 \cdot I}} \tag{2}$$

where R is the the gas constant, T is the temperature, F is the Faraday number, I is the ionic strength of 0.9% NaCl, and $\varepsilon\varepsilon_0$ is the permeability of the electric medium.

3. Results

The experimental values of the surface charge densities of erythrocyte and thrombocyte membranes (control and fatal ethyl alcohol poisoning groups) versus pH are presented in Figures 1 and 2, respectively. We determined the surface charge density values based on experimental electrophoretic mobility data using Equation (1). All measurements were carried out at several pH values, using the basic electrolyte (0.155 M NaCl).

Figure 1 shows the surface charge densities of the control and fatal ethyl alcohol poisoning erythrocyte groups versus pH values. We observed a decrease in the positive charge and an increase in the erythrocytes membrane's negative charge after death by ethyl alcohol poisoning compared to control erythrocytes. A slight shift of the isoelectric point of membranes to lower pH values is visible.

Figure 2 shows the effect of pH on the surface charge density for the control and fatal ethyl alcohol poisoning thrombocyte groups.

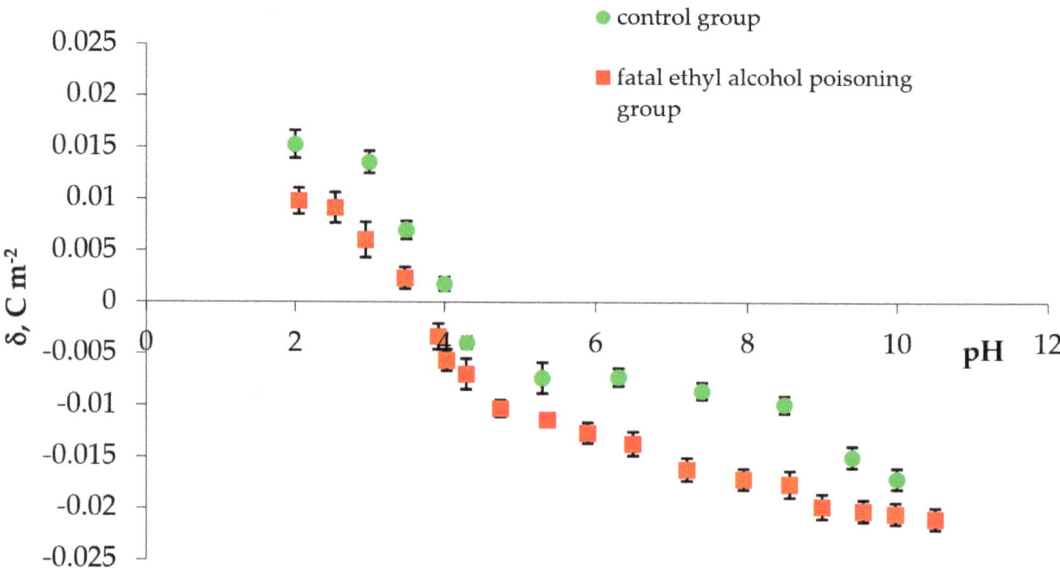

Figure 1. The experimental surface charge density of erythrocytes vs. pH of the electrolyte solution ●—control, ■—fatal ethyl alcohol poisoning.

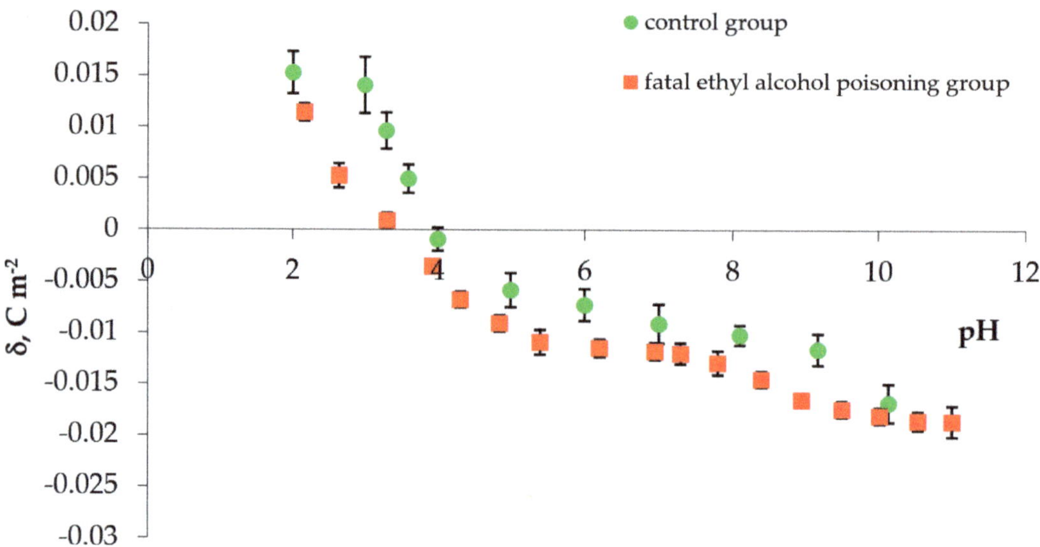

Figure 2. The experimental surface charge density of thrombocytes vs. pH of the electrolyte solution: ●—control, ■—fatal ethyl alcohol poisoning.

Fatal ethyl alcohol poisoning caused a decrease in the positive membrane charge and increased the thrombocyte group's negative charge compared with the control. Also, we observed a shift of the isoelectric point of the membrane to lower pH values.

Tables 1 and 2 show the isoelectric point values and the surface charge densities of the erythrocyte and thrombocyte membranes, as determined by microelectrophoresis. We express the summarized results as the mean value with the designated standard deviation. We performed the analysis using standard statistical analysis.

Table 1. The isoelectric point and surface charge density values for red blood cells (RBCs) (control and fatal ethyl alcohol poisoning).

Examined Groups	Isoelectric Point	Surface Charge Density [10^{-2} C m^{-2}]	
		pH ~ 3	pH ~ 10
control	4.05	1.357 ± 0.113	-1.710 ± 0.109
fatal ethyl alcohol poisoning	3.92	0.978 ± 0.068	-2.300 ± 0.106

Table 2. The surface charge density and isoelectric point values for thrombocytes (control and fatal ethyl alcohol poisoning).

Examined Groups	Isoelectric Point	Surface Charge Density [10^{-2} C m^{-2}]	
		pH ~ 3	pH ~ 10
control	4.20	1.407 ± 0.212	-1.681 ± 0.178
fatal ethyl alcohol poisoning	3.31	1.141 ± 0.138	-1.858 ± 0.155

4. Discussion

The natural cell lipid bilayer is a complex and dynamic protein-lipid structure. The complexity of biological membranes determines many equilibria, both between the membrane components and between cell membranes and the surrounding ions. Particularly important to the cell's functioning are the equilibria between the membrane and the surrounding aqueous solution. The ions present in the solution, adsorbing on the membrane, modulate many of its physicochemical and electrical properties, such as electric charge. Due to the lack of literature data on the effect of fatal poisoning with ethyl alcohol on the equilibria between blood cell membranes and their surroundings, we adopted the theoretical model proposed by Dobrzyńska et al. [32] (presented in full detail in [24]) to describe these equilibria. The given model allowed us to verify the theoretical data with experimental data.

Equations (1)–(7) describe the assumptions of the presented model. H^+, OH^-, Na^+, and Cl^- ions from the electrolyte solution adsorb at the erythrocytes' and thrombocytes' surface. Two of the four equilibria described are associated with positive groups (phospholipids or proteins and sodium or hydrogen ions). The other two are associated with negative groups on the phospholipids' or proteins' surface and hydroxide or chloride ions. The equations of adsorption of H^+, OH^-, Na^+, and Cl^- ions on functional groups located on the membrane surface are presented in Equations (1)–(4):

$$A^- + H^+ \Leftrightarrow AH \tag{3}$$

$$A^- + Na^+ \Leftrightarrow ANa \tag{4}$$

$$B^+ + OH^- \Leftrightarrow BOH \tag{5}$$

$$B^+ + Cl^- \Leftrightarrow BCl \tag{6}$$

$$c_A = a_{A^-} + a_{AH} + a_{ANa} \tag{7}$$

$$c_B = a_{B^+} + a_{BOH} + a_{BCl} \tag{8}$$

$$\delta = \left(a_{B^+} - a_{A^-} \right) \cdot F \tag{9}$$

where a_{AH}, a_{ANa}, a_{A^-}, a_{BOH}, a_{BCl}, and a_{B^+} are the surface concentrations of the corresponding groups on the membrane surface; a_{H^+}, a_{Na^+}, a_{OH^-}, and a_{Cl^-} are the volume concentrations of solution ions; c_A is the total surface concentration of the membrane acidic groups;

c_B is the total surface concentration of the membrane basic groups, $F = 96,487 \left[\frac{C}{mol}\right]$ is the Faraday constant; and δ is the surface charge density.

Final equations [24,32]:

- The equation for determining the membrane surface charge density:

$$\frac{\delta}{F} = \frac{c_B}{1 + K_{BOH}a_{OH^-} + K_{BCl}a_{Cl^-}} - \frac{c_A}{1 + K_{AH}a_{H^+} + K_{ANa}a_{Na^+}} \tag{10}$$

- Linear equations—simplifications of Equation (8)—for high (Equation (9)) and low (Equation (10)) concentrations of hydrogen ions:

$$\frac{\delta a_{H^+}}{F} = \frac{c_B}{1 + K_{BCl}a_{Cl^-}}a_{H^+} - \left(\frac{c_B K_{BOH}K_W}{(1 + K_{BCl}a_{Cl^-})^2} + \frac{c_A}{K_{AH}}\right) \tag{11}$$

$$\frac{\delta}{Fa_{H^+}} = -\left(\frac{c_A}{1 + K_{ANa}a_{Na^+}}\right)\frac{1}{a_{H^+}} + \left(\frac{c_B}{K_{BOH}K_W} + \frac{c_A K_{AH}}{(1 + K_{ANa}a_{Na^+})^2}\right) \tag{12}$$

where K_{AH}, K_{ANa}, K_{BOH}, and K_{BCl} are association constants.

The above linear function coefficients can be easily determined using the linear regression method and then used to calculate the membrane parameters (Equations (9) and (10)). The necessary parameters are the total concentrations of functional acidic (c_A) and basic (c_B) groups on the blood cell surfaces and their average association constants with hydrogen (K_{AH}) and hydroxyl (K_{BOH}) ions. Determination of all the necessary parameters was possible only based on the assumption that the association constant values K_{ANa} and K_{BCl} are the same as the values obtained for phosphatidylcholine liposomes. The K_{ANa} and K_{BCl} values of the surface groups of phosphatidylcholine with sodium and chloride ions were reported previously and amount to 0.230 and 0.076 [m^3/mol], respectively [18]. To calculate the theoretical values of the erythrocyte and thrombocyte membranes' surface charge density, the values of c_A, c_B, K_{AH}, and K_{BOH} were determined and substituted into Equation (8).

The comparison of the experimental (calculated from Equation (11)) and theoretical (calculated from Equation (8)) surface charge density values of the erythrocyte and thrombocyte membranes versus pH are presented in Figures 3 and 4 (the points indicate the experimental data and the curves indicate theoretical data).

The following conclusion can be drawn from the curves presented in Figures 3 and 4: There is an agreement between the theoretical and experimental values in the pH range of 2–7. Above pH 7, the theoretical curves differ from the experimental points. Variations at pH above 7 may be caused by interactions between the functional groups of the cell membranes of blood morphotic elements. The mathematical model we proposed takes into account only the equilibrium of the membrane surface with electrolyte ions.

In Tables 3 and 4, we summarize the obtained values of the parameters characterizing the equilibria between the erythrocyte and thrombocyte surfaces. The presented results were analyzed using standard statistical analysis and are expressed as means with standard deviations. The effect of fatal alcohol poisoning on erythrocyte and thrombocyte membranes caused changes in the values of the calculated parameters (c_A, c_B, K_{AH}, and K_{BOH}) (Tables 3 and 4).

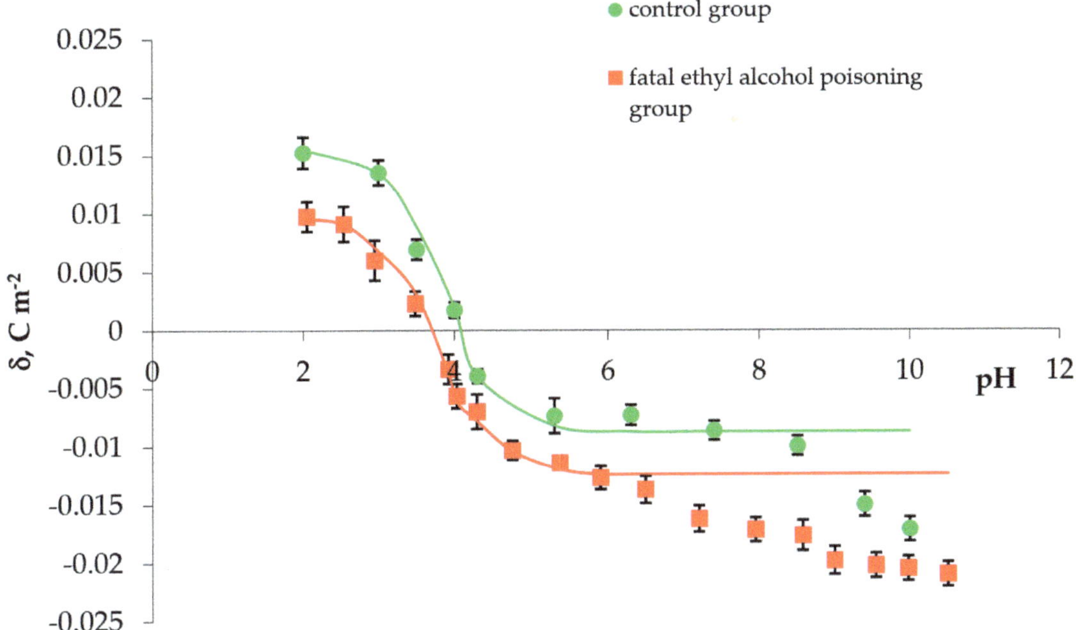

Figure 3. A comparison of experimental and theoretical surface charge density values of erythrocytes vs. pH of the electrolyte solution: ● control, ■ fatal ethyl alcohol poisoning. Points represent the experimental values and curves represent theoretical values.

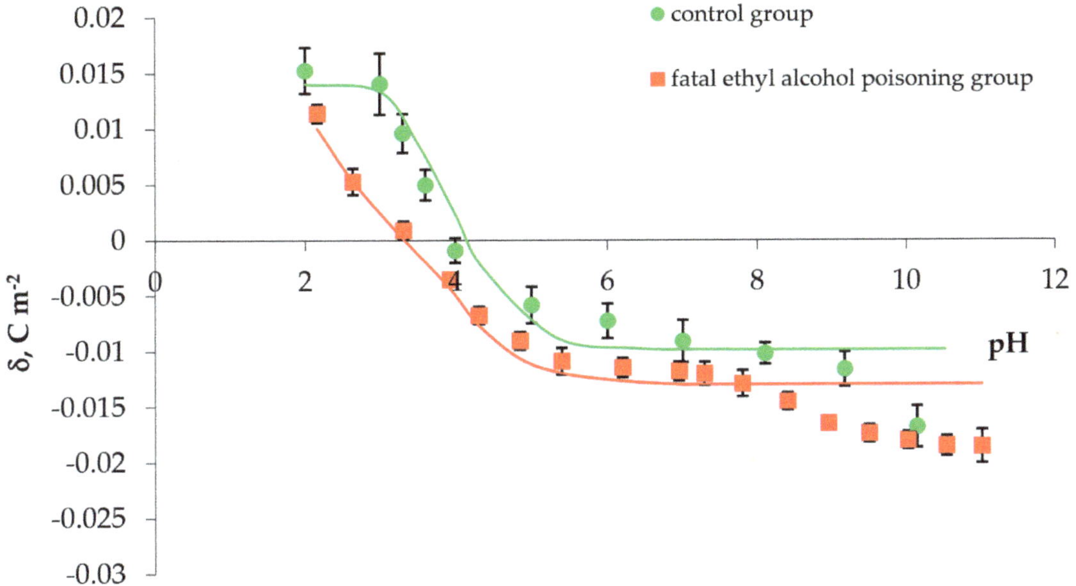

Figure 4. A comparison of experimental and theoretical surface charge density values of thrombocytes vs. pH of the electrolyte solution: ● control, ■ fatal ethyl alcohol poisoning. Points represent the experimental values and curves represent theoretical values.

Table 3. The acidic and basic total concentrations of functional groups of erythrocytes and their association constants with H^+ and OH^- ions.

Groups	Parameters			
	c_A [10^{-6} mol/m^2]	c_B [10^{-6} mol/m^2]	K_{AH} [10^2 m^3/mol]	K_{BOH} [10^7 m^3/mol]
control	7.11 ± 0.44	1.61 ± 0.51	3.44 ± 1.11	3.56 ± 0.81
fatal ethyl alcohol poisoning	1.16 ± 0.08	3.22 ± 0.91	0.38 ± 0.04	2.12 ± 0.26

Table 4. The total acid and base concentrations of the functional groups of thrombocytes and their association constants with H^+ and OH^- ions.

Groups	Parameters			
	c_A [10^{-6} mol/m^2]	c_B [10^{-6} mol/m^2]	K_{AH} [10^2 m^3/mol]	K_{BOH} [10^7 m^3/mol]
control	3.76 ± 0.81	1.21 ± 0.23	2.88 ± 1.58	2.11 ± 0.79
fatal ethyl alcohol poisoning	5.56 ± 0.33	0.17 ± 0.03	3.47 ± 0.65	5.21 ± 0.85

In this study, the c_A, K_{AH}, and K_{BOH} values for an erythrocyte cell membrane were affected by ethyl alcohol and were smaller than the same parameters assayed in unmodified cells; only c_B values were higher (approximately double, Table 3).

In the case of a thrombocyte cell membrane, the c_A, K_{AH}, and K_{BOH} parameters were affected by ethyl alcohol and were higher than the same parameters assayed in unmodified cells. Only c_B values were smaller, by about seven times (Table 4).

Alcohol intoxication causes changes in the amounts of both phospholipids and integral proteins of erythrocyte membranes [17], resulting in the appearance of new negatively and positively charged functional groups. Therefore, variations in the kind and number of these groups result in changes in the parameters describing equilibria in cell membranes: c_A, c_B, K_{AH}, and K_{BOH}. It is believed that other mechanisms can also influence the above parameters, such as the movement of molecules forming membranes between monolayers (the flip-flop phenomenon) or the appearance of receptor proteins or their components on the surface of cell membranes, which, as a result of stopped life processes, no longer perceive signals from the external environment and remain in an inactive form. Scheidt and Huster [33] demonstrated that ethyl alcohol partitions the lipid–water interface of the phospholipid bilayer by forming hydrogen bonds with lipid molecules and, also, by the hydrophobic effect. Rowe showed that at a higher ethanol concentration, a significant reduction (up to 30%) in phosphatidylcholine lipid membrane thickness is observed as it transitions into the interdigitated phase [34]. In natural lipid membranes, ethanol-induced membrane perturbations may have many possible effects. A considerable reduction in membrane thickness caused by lipid interdigitation would likely profoundly affect membrane protein function and conformation [35]. Membrane thickness changes can result in the exposure of hydrophobic amino acid residuum in integral proteins of the membrane and an occurrence like a hydrophobic mismatch, leading to membrane protein aggregation [36] and possibly producing a conformational change in the membrane protein [37]. Zeng et al. [38] demonstrated that ethanol-induced fluidization and interdigitation of the lipid bilayer lead to an increase in the membrane's ion permeability. Lee [39] examined the effects of ethanol exposure on human erythrocytes using quantitative phase imaging techniques at the level of individual cells. The author demonstrated that erythrocytes exposed to ethanol (at concentrations of 0.1 and 0.3% v/v) exhibited cell sphericities higher than those of normal cells. Bulle and co-workers [40] studied the association between erythrocyte membrane alterations and hemolysis in chronic alcoholics. They showed that frequent alcohol consumption increases oxidative/nitrosative stress. The result is a change in the lipid composition of the erythrocyte membrane and the protein content, resulting from increased hemolysis.

There are no data on ethyl alcohol's effect on blood platelets' electrical properties in the literature. Slowed blood flow in ethyl alcohol poisoning promotes platelet activation, leading to changes in the exposure of phospholipids (phosphatidylserine and phosphatidylethanolamine) in the clotting process to the surface of the cell membrane. Since phospholipids are endowed with an electric charge, we believe that their presence in the membrane will result in changes in the surface charge density and, consequently, the determined parameters (c_A, c_B, K_{AH}, and K_{BOH}).

5. Conclusions

Based on numerous literature reports, it can be said that ethyl alcohol shows biophysical effects on lipid membranes. In this paper, both experimental and theoretical data demonstrated that the electrical properties of blood cell membranes (erythrocytes and thrombocytes) are changed upon fatal alcohol poisoning. We observed changes in all determined parameters, c_A, c_B, K_{AH}, and K_{BOH}. Interactions among membrane components (lipids and proteins) and between these components and their surroundings lead to variations in the number and kind of functional groups present, which results in changes in c_A and c_B and in the values of association constants. Also, sialic acid, a component of glycoproteins and glycolipids, may influence all these parameters in fatal ethyl alcohol poisoning. Understanding some of the physical and chemical processes occurring in the human body after death is essential in describing the pathophysiology of fatal ethyl alcohol poisoning deaths. A quantitative description of the phenomena occurring on the blood cell membrane surfaces and the obtained parameters characterizing the existing equilibria (c_A, c_B, K_{AH}, and K_{BOH}) may provide essential information for forensic medicine. We want to emphasize that our theoretical descriptions are innovative and the research carried out is quite preliminary. Still, because the obtained data may help solve many forensic medicine problems, there is no doubt that there is a necessity for continuation of this research.

Author Contributions: Conceptualization, M.S. and A.D.P.; methodology, J.K. and A.D.P.; validation, M.S., J.K., and A.D.P.; formal analysis, J.K. and A.D.P.; writing—original draft preparation, M.S., J.K., A.N.-J., A.D.P.; writing—review and editing, M.S., J.K., A.N.-J., and A.D.P.; visualization, M.S. and A.D.P.; supervision, A.D.P. All authors have read and agreed to the published version of the manuscript.

Funding: This research received no external funding.

Institutional Review Board Statement: The study was conducted according to the guidelines of the Declaration of Helsinki, and approved by the The Ethics Review Board of the Medical University of Bialystok approved the conducted research (No. R-I-002/533/2010).

Informed Consent Statement: Not applicable.

Data Availability Statement: The data presented in this study are available on request from the corresponding author.

Acknowledgments: The Zetasizer Nano ZS apparatus was funded by the European Funds for Regional Development and the National Funds of the Ministry of Science and Higher Education as part of the Operational Program Development of Eastern Poland 2007–2013 [POPW.01.03.00-20-044/11].

Conflicts of Interest: The authors declare no conflict of interest.

References

1. Ballard, H.S. The hematological complications of alcoholism. *Alcohol. Clin. Exp. Res.* **1989**, *13*, 706–720. [CrossRef] [PubMed]
2. Ballard, H.S. Alcohol, bone marrow, and blood. *Alcohol. Health Res. World.* **1993**, *17*, 310–315.
3. Stouten, K.; Riedl, J.A.; Droogendijk, J.; Castel, R.; van Rosmalen, J.; van Houten, R.J.; Berendes, P.; Sonneveld, P.; Levin, M.D. Prevalence of potential underlying aetiology of macrocytic anaemia in Dutch general practice. *BMC Fam. Pr.* **2016**, *17*, 113. [CrossRef]

4. Thoma, E.; Bitri, S.; Mucaj, K.; Tahiri, A.; Pano, I. Changes of Some Blood Count Variables in Correlation with the Time of Alcohol Abuse. *J. Addict. Res.* **2015**, *6*, 2. [CrossRef]

5. Marumo, M.; Wakabayashi, I. Sensitivity of thrombin-induced platelet aggregation to inhibition by ethanol. *Clin. Chim. Acta* **2009**, *402*, 156–159. [CrossRef]

6. Ballard, H.S. *Alcohol Abuse in Guide to Anemia*, 2nd ed.; Garrison, C.D., Ed.; Cumberland House: Naperville, IL, USA, 2009; p. 165.

7. De Pietri, L.; Bianchini, M.; Montalti, R.; De Maria, N.; Di Maira, T.; Begliomini, B.; Gerunda, G.E.; di Benedetto, F.; Garcia-Tsao, G.; Villa, E. Thrombelastography-guided blood product use before invasive procedures in cirrhosis with severe coagulopathy: A randomized, controlled trial. *Hepatology* **2016**, *63*, 566–573. [CrossRef] [PubMed]

8. Gurtovenko, A.A.; Anwar, J. Interaction of Ethanol with Biological Membranes: The Formation of Non-bilayer Structures within the Membrane Interior and their Significance. *J. Phys. Chem. B* **2009**, *113*, 1983–1992. [CrossRef]

9. Sonmez, M.; Ince, H.Y.; Yalcin, O.; Ajdžanović, V.; Spasojević, I.; Meiselman, H.J.; Baskurt, O.K. The effect of alcohols on red blood cell mechanical properties and membrane fluidity depends on their molecular size. *PLoS ONE* **2013**, *8*, e76579. [CrossRef]

10. Dora, M.D.; Goldstein, B. Effect of alcohol on cellular membranes. *Ann. Emerg. Med.* **1986**, *15*, 1013–1018.

11. Chin, J.H.; Goldstein, B.D. Effects of low concentrations of ethanol on the fluidity of spin-labeled erythrocyte and brain membranes. *Mol. Pharm.* **1977**, *13*, 435–441.

12. Wood, W.G.; Lahiri, S.; Gorka, C.; Armbrecht, H.J.; Strong, R. In vitro effects of ethanol on erythrocyte membrane fluidity of alcoholic patients: An electron spin resonance study. *Alcohol. Clin. Exp. Res.* **1987**, *11*, 332–335. [CrossRef]

13. Beaug, F.; Gallay, J.; Stibler, H.; Borg, S. Alcohol abuse increases the lipid structural order in human erythrocyte membranes: A steady-state and time-resolved anisotropy study. *Biochem. Pharmacol.* **1988**, *37*, 3823–3828. [CrossRef]

14. Stibler, H.; Beauge, F.; Leguicher, A.; Borg, S. Biophysical and biochemical alterations in erythrocyte membranes from chronic alcoholics. *Scand. J. Clin. Lab. Investig.* **1991**, *51*, 309–319. [CrossRef]

15. Hrelia, S.; Lercker, G.; Biagi, P.; Bordoni, A.; Stefanini, F.; Zunarelli, P.; Rossi, C.A. Effect of ethanol intake on human erythrocyte membrane fluidity and lipid composition. *Biochem. Int.* **1986**, *12*, 741–750.

16. Dobrzynska, I.; Szachowicz-Petelska, B.; Skrzydlewska, E.; Figaszewski, Z.A. Changes in electric charge and phospholipids composition in erythrocyte membrane of ethanol-poisoned rats after administration of teas. *Acta Pol. Pharm.* **2004**, *61*, 483–487. [PubMed]

17. Szachowicz-Petelska, B.; Dobrzynska, I.; Skrzydlewska, E.; Figaszewski, Z. Influence of green tea on surface charge density and phospholipids composition of erythrocytes membrane in ethanol intoxicated rats. *Cell Biol. Toxicol.* **2005**, *21*, 61–70. [CrossRef]

18. Szachowicz-Petelska, B.; Dobrzynska, I.; Skrzydlewska, E.; Figaszewski, Z. Protective effect of blackcurrant on liver cell membrane of rats intoxicated with ethanol. *J. Membr. Biol.* **2012**, *245*, 191–200. [CrossRef] [PubMed]

19. Chachaj-Brekiesz, A.; Kobierski, J.; Wnętrzak, A.; Dynarowicz-Latka, P. Electrical Properties of Membrane Phospholipids in Langmuir Monolayers. *Membranes* **2021**, *11*, 53. [CrossRef]

20. Naumowicz, M.; Zając, M.; Kusaczuk, M.; Gál, M.; Kotyńska, J. Electrophoretic Light Scattering and Electrochemical Impedance Spectroscopy Studies of Lipid Bilayers Modified by Cinnamic Acid and Its Hydroxyl Derivatives. *Membranes* **2020**, *10*, 343. [CrossRef] [PubMed]

21. Dziubak, D.; Strzelak, K.; Sek, S. Electrochemical Properties of Lipid Membranes Self-Assembled from Bicelles. *Membranes* **2021**, *11*, 11. [CrossRef]

22. Fernandes, H.P.; Cesar, C.L.; de Lourdes Barjas-Castro, M. Electrical properties of the red blood cell membrane and immunohematological investigation. *Rev. Bras. Hematol. Hemoter.* **2011**, *33*, 297–301. [CrossRef] [PubMed]

23. Liu, J.; Qiang, Y.; Alvarez, O.; Du, E. Electrical Impedance Characterization of Erythrocyte Response to Cyclic Hypoxia in Sickle Cell Disease. *ACS Sens.* **2019**, *4*, 1783–1790. [CrossRef] [PubMed]

24. Kotyńska, J.; Petelska, A.D.; Szeremeta, M.; Niemcunowicz-Janica, A.; Figaszewski, Z.A. Changes in surface charge density of blood cells after sudden unexpected death. *J. Membr. Biol.* **2012**, *245*, 185–190. [CrossRef]

25. Szeremeta, M.; Petelska, A.D.; Kotyńska, J.; Niemcunowicz-Janica, A.; Figaszewski, Z.A. The effect of fatal carbon monoxide poisoning on the surface charge of blood cells. *J. Membr. Biol.* **2013**, *246*, 717–722. [CrossRef] [PubMed]

26. Petelska, A.D.; Kotyńska, J.; Figaszewski, Z.A. The effect of fatal carbon monoxide poisoning on the equilibria between cell membranes and the electrolyte solution. *J. Membr. Biol.* **2015**, *248*, 157–161. [CrossRef] [PubMed]

27. Szeremeta, M.; Petelska, A.D.; Kotyńska, J.; Pepiński, W.; Naumowicz, M.; Figaszewski, Z.A.; Niemcunowicz-Janica, A. Changes in surface charge density of blood cells in fatal accidental hypothermia. *J. Membr. Biol.* **2015**, *248*, 1175–1180. [CrossRef]

28. Petelska, A.D.; Kotyńska, J.; Naumowicz, M.; Figaszewski, Z.A. Equilibria between cell membranes and electrolyte solution: Effect of fatal accidental hypothermia. *J. Membr. Biol.* **2016**, *249*, 375–380. [CrossRef] [PubMed]

29. Patra, M.; Salonen, E.; Terama, E.; Vattulainen, I.; Faller, R.; Lee, B.W.; Holopainen, J.; Karttunen, M. Under the Influence of Alcohol: The Effect of Ethanol and Methanol on Lipid Bilayers. *Biophys. J.* **2006**, *90*, 1121–1135. [CrossRef]

30. Alexander, A.E.; Johnson, P. *Colloid Science*; Clarendon Press: Oxford, UK, 1949.

31. Barrow, G.M. *Physical Chemistry*; McGraw-Hill Inc: New York, NY, USA, 1996.

32. Dobrzyńska, I.; Skrzydlewska, E.; Figaszewski, Z. Parameters characterizing acid-base equilibria between cell membrane and solution and their application to monitoring the effect of various factors on the membrane. *Bioelectrochemistry* **2006**, *69*, 142–147. [CrossRef] [PubMed]

33. Scheidt, H.A.; Huster, D. The interaction of small molecules with phospholipid membranes studied by 1 H NOESY NMR under magic-angle spinning. *Acta Pharmacol. Sin.* **2008**, *29*, 35–49. [CrossRef] [PubMed]
34. Rowe, E.S. Effects of ethanol on membrane lipids. In *Alcohol and Neurobiology: Receptors, Membranes, and Channels*, 1st ed.; Watson, R.R., Ed.; CRC Press: Boca Raton, FL, USA, 1992; pp. 239–268.
35. Henderson, C.M.; Block, D.E. Examining the Role of Membrane Lipid Composition in Determining the Ethanol Tolerance of Saccharomyces cerevisiae. *Appl. Environ. Microbiol.* **2014**, *80*, 2966–2972. [CrossRef]
36. Lee, S.Y.; Park, H.J.; Best-Popescu, C.; Jang, S.; Park, Y.K. The Effects of Ethanol on the Morphological and Biochemical Properties of Individual Human Red Blood Cells. *PLoS ONE* **2015**, *10*, e0145327. [CrossRef] [PubMed]
37. Killian, J.A. Hydrophobic mismatch between proteins and lipids in membranes. *Biochim. Biophys. Acta.* **1998**, *1376*, 401–416. [CrossRef]
38. Zeng, J.; Smith, K.E.; Chong, P.L. Effects of alcohol-induced lipid interdigitation on proton permeability in L-α-dipalmitoylphosphatidylcholine vesicles. *Biophys. J.* **1993**, *65*, 1404–1414. [CrossRef]
39. Lee, A.G. How lipids affect the activities of integral membrane proteins. *Biochim. Biophys. Acta* **2004**, *1666*, 62–87. [CrossRef]
40. Bulle, S.; Reddy, V.D.; Padmavathi, P.; Maturu, P.; Puvvada, P.K.; Nallanchakravarthula, V. Association between alcohol-induced erythrocyte membrane alterations and hemolysis in chronic alcoholic. *J. Clin. Biochem. Nutr.* **2017**, *60*, 63–69. [CrossRef] [PubMed]

 *membranes*

Article

Water Pores in Planar Lipid Bilayers at Fast and Slow Rise of Transmembrane Voltage

Alenka Maček Lebar [1,*,†], **Damijan Miklavčič** [1], **Malgorzata Kotulska** [2], **Peter Kramar** [1,*,†]

1. Faculty of Electrical Engineering, University of Ljubljana, 1000 Ljubljana, Slovenia; damijan.miklavcic@fe.uni-lj.si
2. Department of Biomedical Engineering, Faculty of Fundamental Problems of Technology, Wroclaw University of Science and Technology, 50-370 Wrocław, Poland; malgorzata.kotulska@pwr.edu.pl
* Correspondence: alenka.macek.lebar@fe.uni-lj.si (A.M.L.); peter.kramar@fe.uni-lj.si.com (P.K.); Tel.: +386-14768770 (A.M.L.); +386-14768290 (P.K.)
† These authors contributed equally to this work.

Abstract: Basic understanding of the barrier properties of biological membranes can be obtained by studying model systems, such as planar lipid bilayers. Here, we study water pores in planar lipid bilayers in the presence of transmembrane voltage. Planar lipid bilayers were exposed to fast and slow linearly increasing voltage and current signals. We measured the capacitance, breakdown voltage, and rupture time of planar lipid bilayers composed of 1-pamitoyl 2-oleoyl phosphatidylcholine (POPC), 1-pamitoyl 2-oleoyl phosphatidylserine (POPS), and a mixture of both lipids in a 1:1 ratio. Based on the measurements, we evaluated the change in the capacitance of the planar lipid bilayer corresponding to water pores, the radius of water pores at membrane rupture, and the fraction of the area of the planar lipid bilayer occupied by water pores. The estimated pore radii determining the rupture of the planar lipid bilayer upon fast build-up of transmembrane voltage are 0.101 nm, 0.110 nm, and 0.106 nm for membranes composed of POPC, POPS, and POPC:POPS, respectively. The fraction of the surface occupied by water pores at the moment of rupture of the planar lipid bilayer is in the range of 0.1–1.8%.

Keywords: planar lipid bilayer; capacitance; voltage breakdown; water pores; hydrophobic pores

Citation: Maček Lebar, A.; Miklavčič, D.; Kotulska, M.; Kramar, P. Water Pores in Planar Lipid Bilayers at Fast and Slow Rise of Transmembrane Voltage. *Membranes* **2021**, *11*, 263. https://doi.org/10.3390/membranes11040263

Academic Editors: Francisco Monroy and Monika Naumowicz

Received: 7 March 2021
Accepted: 29 March 2021
Published: 5 April 2021

1. Introduction

Biological membranes, the barriers that envelope the cell and its inner organelles, play an important role in the physiology and functionality of cells. Although biological membranes are composed of lipids, proteins, and small amounts of carbohydrates, the barrier function is assured by the thin lipid bilayer. Basic understanding of barrier properties of biological membranes can be obtained by investigating model systems, such as artificial liposomes or vesicles, which mimic the geometry and size of the cell membrane but are devoid of ion channels and the multitude of other embedded components, commonly present in the cell membrane. The simplest model of a cell membrane patch is a planar lipid bilayer. The chemical composition of the planar lipid bilayer can be chosen in advance and is therefore well defined. However, to mimic a non-curved fragment of the cell membrane, the planar lipid bilayer should separate two electrolytes. Therefore, the planar lipid bilayer is usually vertically formed across a small aperture in a hydrophobic partition that separates two compartments filled with electrolytes [1]. Visual observation of the planar lipid bilayer in such an experimental system is limited. However, electrodes immersed in the electrolyte permit measurements of electric parameters of the planar lipid bilayer. As a planar lipid bilayer can be electrically considered as a non-perfect capacitor, the capacitance C of an ideal capacitor and the resistance R of a resistor in parallel both describe planar lipid bilayer's characteristic [2,3]. Voltage-controlled and current-controlled methods enable observing electrical properties of the planar lipid bilayer and its structural changes that are reflected in the electrical characteristics [4–7].

The capacitance C is the parameter considered as the best tool for probing the stability and proper formation of planar lipid bilayer and for this reason, it is measured for each planar lipid bilayer studied with electrical methods, even when other properties are in the focus of investigation. For comparison between different studies, the measured value of the capacitance is normalized by the area of planar lipid bilayer to determine the specific capacitance c [8].

An exposure of planar lipid bilayer to external electrical stimulus and formation of transmembrane voltage alter both electrical parameters of planar lipid bilayer model, R and C. Molecular dynamics (MD) simulations have shown that small so-called water fingers are formed on both sides of the planar lipid bilayer in the presence of transmembrane voltage and create water or hydrophobic pores [9,10]. The lipids adjacent to the water molecules inside the pores start reorienting their polar headgroups toward water molecules and stabilizing the pores into hydrophilic state, which allows more water and ions to enter the pores. These conductive paths reduce R of the planar lipid bilayer. At low transmembrane voltages, when pores are not created yet, a capacitance increase can be observed. An electrostrictive thinning of the planar lipid bilayer has been suggested as the mechanism responsible for this phenomenon [11,12]. The capacitance of the planar lipid bilayer depends on applied voltage U according to the equation

$$C(U) = C_0[1 + \alpha U^2], \tag{1}$$

where C_0 is the planar lipid bilayer capacitance at zero transmembrane voltage, and α is the proportionality coefficient. Experiments revealed that its value is around $0.02\ V^{-2}$ [11], while using numerical models it was assessed to be in the range of 0.053–$0.082\ V^{-2}$ [13]. However, at higher transmembrane voltages, appearance of water and hydrophilic pores leads to reduction of the planar lipid bilayer capacitance, due to significant difference in dielectric constant values of lipid bilayer and water [14,15]. A long-lasting exposure to strong electric stimulus causes irreversible damage to a planar lipid bilayer. Electrical measurements permit determination of the breakdown voltage value U_{br}.

Hydrophobic pores formed in planar lipid bilayer were first hypothesized decades ago in Abidor's theory of electroporation [16] and as water pores successfully modeled by molecular dynamic simulation [17]. Since such water pores are very small transient structures, they cannot be directly observed. As far as we know, the only study that associated experimental data with the existence of hydrophobic pores in the planar lipid bilayers were reported by Anosov et al. [18]. The experimentally observed increase in the variance of membrane current, that was attributed to small current fluctuations due to membrane capacitance changes, was connected to the number of hydrophobic pores in the membrane during the phase transition. Even experimental information about conducting hydrophilic pores and their characteristics comes only from indirect measurements, like current fluctuations measurements [19], voltage fluctuations measurements [20,21], or observing voltage drops [22]. Recently, Akimov et al. [15,23] introduced an improved theory on the pore formation, which allows modeling continuous trajectories of pore formation, both in the absence and presence of stress conditions. The theory nicely explains the complex behavior of pores during their formation under lateral tension, even in various loading rates regimes that were studied in giant unilamellar vesicles (GUV) aspiration experiments [24]. Namely, at low loading rates, GUV membrane ruptured at small tensions and, according to theory, membrane breakdown is governed by the pore expansion. On the other hand, at high loading rates, GUV membrane rupture occurs at high values of tensions and is theoretically limited by water pore formation, which is lipid type (curvature) dependent. Akimov's theory predicts that the planar lipid bilayer, exposed to the electric field, behaves similarly as in the presence of lateral tension.

To explore if different membrane rupture mechanisms can be predicted also from experiments in which transmembrane voltage is raised at various rates, we exposed planar lipid bilayers to linearly rising voltage or current signals of different slopes. Planar lipid bilayers were formed using two types of lipid molecules: 1-pamitoyl 2-oleoyl phosphatidylcholine (POPC), lipid molecules with a zwitterionic head group and almost zero spontaneous curvature, negatively charged 1-pamitoyl 2-oleoyl phosphatidylserine (POPS) molecules with negative spontaneous curvature, as well as mixture of both lipid types in a 1:1 ratio. For each planar lipid bilayer composition, C was measured using two different methods—discharge pulse [4,25] and voltage to period converter [26,27]. Changing rates of transmembrane voltage rise and measuring U_{br} and time at which planar lipid bilayer rupture occurs t_{br}, we evaluated changes in C that correspond to water pores, estimated water pore radius at the rupture moment, and calculated the fraction of planar lipid bilayer area that is occupied by water pores.

2. Materials and Methods

2.1. Chemicals

The lipids 1-pamitoyl 2-oleoyl phosphatidylcholine (POPC) and 1-pamitoyl 2-oleoyl phosphatidylserine (POPS) were purchased in powder form (Avanti Polar-Lipids Inc., Birmingham, AL, USA). A solution of 10 mg/mL of each lipid was prepared in a 9:1 mixture of hexane and ethanol. A mixture of both lipids in a 1:1 ratio was prepared by pipetting the same amount of both lipid solutions in a small plastic tube and mixing by repeated pipetting immediately before the experiments. The mixture of hexadecane and pentane in the ratio of 3:7 was used for torus formation. The electrolyte was prepared from 0.1 M KCl and 0.01 M Hepes in the ratio of 1:1. A few drops of 1 M NaOH were added to obtain a pH of 7.4.

2.2. Experimental Setups

Planar lipid bilayers were formed by the Montal–Muller method [28] across a circular hole 117 μm in diameter (d). The hole was in a thin Teflon sheet sandwiched between two reservoirs in the Teflon chamber. The glass window in the Teflon chamber allowed easy viewing of the hole during the formation of the planar lipid bilayer using stereo microscope with led illumination. Both sides of the hole in the Teflon sheet were first treated with 1 μL of the lipid solution. After the lipid solvent had completely evaporated, the hole was treated on each side with 1.5 μL of the hexadecane and pentane mixture. Both reservoirs were filled with electrolyte just below the hole. A drop of 2 μL of the lipid solution was carefully applied to the electrolyte surface in each reservoir and allowed to spread for at least 15 min. To form a planar lipid bilayer, the electrolyte level in both reservoirs was raised to the same height just above the hole (Figure 1A). After a planar lipid bilayer was ruptured, a new one was formed by lowering the electrolyte levels below the hole, and then raising them again. All experiments were performed at room temperature (24 °C).

The voltage-controlled measurement system [29] consisted of a signal generator, two Ag-AgCl electrodes (E-205, IVM, Healdsburg, CA, USA) immersed in the electrolyte, one on each side of the planar lipid bilayer, and amplifiers connected to an oscilloscope for measuring transmembrane current and voltage. The electrodes were used to apply the transmembrane voltage and measure the transmembrane current. The transmembrane voltage was measured using the LeCroy differential amplifier 1822 and custom made ampere meter was used for transmembrane current measurements. Both signals were stored in Matlab format using LeCroy Waverunner-2 354M oscilloscope and analyzed afterwards.

The current-controlled experiments were performed using a potentiostat-galvanostat measurement system controlled by a computer as described in [20]. Four Ag-AgCl electrodes (E-205, IVM, Healdsburg, CA, USA) immersed in the electrolyte were used, two on each side of the planar lipid bilayer. One pair of electrodes was used to apply current and the other was used as a reference electrodes. The transmembrane voltage measurements were stored on a computer in a raw file and later imported into Matlab for further analysis.

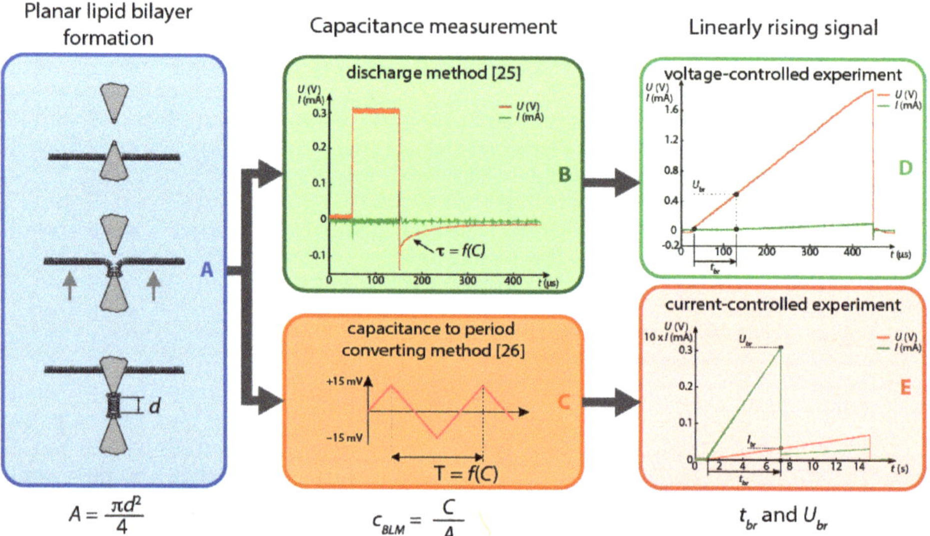

Figure 1. The measurement protocol consisted of formation of the planar lipid bilayer (**A**), capacitance measurement (**B** or **C**), and measurement of the breakdown voltage U_{br} of the planar lipid bilayer (**D** or **E**).

2.3. Measurement Protocols

The measurement protocol consisted of planar lipid bilayer formation (Figure 1A), capacitance measurement (Figure 1B,C) and measurement of the breakdown voltage of the planar lipid bilayer (Figure 1D,E). C, U_{br}, and t_{br} were determined for each planar lipid bilayer. The capacitance C was normalized to the area of the hole ($A = 1.075 \times 10^{-8}$ m^2) to calculate the specific capacitance of the planar lipid bilayer c_{BLM}. When a voltage-controlled measurement system was used, C was determined by the planar lipid bilayer discharge method (Figure 1B) [4]. For current-controlled experiments, C was measured by converting the capacitance to the time period (Figure 1C), which was described in detail by Kalinowski et al. [26].

Transmembrane voltage on planar lipid bilayer was buid-up using a linearly rising signal. In the voltage-controlled experiments, rapid transmembrane voltage build-up was ensured by seven different slopes of the linearly rising voltage k_u: 4.8 kV/s, 5.5 kV/s, 7.8 kV/s, 11.5 16.7 kV/s, 21.6 kV/s, and 48.1 kV/s [25]. At the time t_{br}, when a sudden increase in transmembrane current was detected, U_{br} was measured (Figure 1D). In the current-controlled experiments, a slow build-up of the transmembrane voltage was obtained by five different slopes of the linearly rising current k_i: 0.5 µA/s, 1 µA/s, 4 µA/s, 8 µA/s, and 10 µA/s. At the time t_{br} when a sudden drop of the voltage was detected, we determined U_{br} (Figure 1E). A linearly rising current signal should produce a quadratic shape of the voltage induced on a planar lipid bilayer, but due to the electrical properties of the experimental system setup, the capacitance and resistance of the chamber flattened the voltage curve.

2.4. Experimental Data Analysis

The specific capacitance c_{BLM} was determined for at least 25 planar lipid bilayers of each lipid composition and for each measurement method. Mean value and standard deviation were calculated for each experimental group. Differences in c_{BLM} means between two measurement methods for each lipid composition were determined by t-test. For each lipid composition, at least three measurements of U_{br} were made for each slope of linearly rising voltage and current. To compare U_{br} measured for a given lipid composition at different slopes of the voltage signal, the one-way ANOVA test was used. When comparing the mean breakdown voltages at the same slopes k_u or k_i between lipid bilayers of one component, the unpaired t-test was used. Since the variances of the mean breakdown voltages at slopes $k_u = 7.8$ kV/s and $k_u = 11.5$ kV/s were statistically different, the comparisons were made using Mann–Whitney's test. The difference was considered statistically significant at the values of $p < 0.05$.

Using nonlinear regression, a two-parameter strength–duration curve

$$U = \sqrt[4]{a + \frac{b}{t}} \tag{2}$$

based on the viscoelastic model of Dimitrov [30] and proposed by Sabotin et al. [31] was fitted to experimentally obtained points (t_{br}, U_{br}) measured at different slopes of the linearly rising voltage. The parameter $\sqrt[4]{a}$ is an asymptote of the strength–duration curve corresponding to minimal breakdown voltage U_{brmin} for planar lipid bilayers of a given lipid composition at fast transmembrane voltage build-up. The parameter b determines the inclination of the strength–duration curve. As in the case of strength–duration curve for excitable cell membranes of nerves and muscles, we can define for each lipid composition a kind of a chronaxie time constant $t_c = \frac{b}{15 \cdot a}$ as the minimum time required for a stimulus of twice U_{brmin} to cause rupture of the planar lipid bilayer.

2.5. The Rate of the Planar Lipid Bilayer Capacitance Change at t_{br}

Current $i(t)$ which charges planar lipid bilayer can be represented as

$$i(t) = \frac{u(t)}{R(t)} + \frac{dQ(t)}{dt} = \frac{u(t)}{R(t)} + \frac{d(C(t)u(t))}{dt} = \frac{u(t)}{R(t)} + C(t)\frac{du(t)}{dt} + u(t)\frac{C(t)}{dt}. \tag{3}$$

During voltage-controlled experiments, voltage $u(t)$ rises with time linearly in accordance with the slope k_u. Therefore, Equation (3) changes to

$$i(t) = \frac{k_u}{R(t)}t + C(t)k_u + k_u t\frac{dC(t)}{dt}. \tag{4}$$

$i(t)$ at the moment t_{br} is

$$i(t_{br}) = \frac{k_u}{R(t_{br})}t_{br} + C(t_{br})k_u + k_u t_{br}\frac{d(C(t))}{dt}\bigg|_{t_{br}}. \tag{5}$$

By using experimentally obtained values for t_{br} and $i(t_{br})$, $\frac{d(C(t))}{dt}\big|_{t_{br}}$ can be calculated. We assumed that the changes in C due to electrostrictive thinning are small enough to be neglected. Therefore, for each lipid composition, C measured before exposure of the planar lipid bilayer to linearly rising signal was used in calculations. As a result that we cannot measure R before the membrane breakdown occurs, we calculated it for each lipid composition from current-controlled experiments, performed with the slowest linearly rising current ($k_i = 0.5$ µA) at the moment t_{br}.

We assumed that the last term in Equation (3) is zero, because C has not been changed yet, and the current rises linearly in time according to $i(t) = k_i \cdot t$. Differential equation for transmembrane voltage $u(t)$ was solved using Laplace transform. The solution is

$$u(t) = k_i \cdot R \cdot \left(t - RC + (RC)^2 \cdot e^{-\frac{t}{RC}}\right), \quad t > 0. \tag{6}$$

The equation is valid also at the moment t_{br}. In the case of slowly linearly rising current, the product RC is at least 10 times smaller than t_{br}. Thus, the last summand in the parenthesis in Equation (6) is much smaller than the other two and can be neglected. The Equation (6) can be simplified to

$$C \cdot R^2 - t_{br} \cdot R + \frac{u(t_{br})}{k_i} = 0,$$

(7)

and solved for R.

2.6. Calculation of a Fraction of the Planar Lipid Bilayer That Is Occupied By Pores

In the short interval before t_{br} we can assume that non-conductive water pores are already present in the planar lipid bilayer [14,15,32]. Therefore, C_{br} can be represented as

$$C_{br} = \sum_{i=1}^{n} C_{pi} + \sum_{j=1}^{m} C_{lj},$$

(8)

where $C_{pi} = \varepsilon_p \varepsilon_0 \frac{A_{pi}}{D}$ is the capacitance of the ith water pore and $C_{lj} = \varepsilon_l \varepsilon_0 \frac{A_{lj}}{D}$ capacitance of jth patch of the still intact planar lipid bilayer. ε_p is a dielectric constant of water pores ($\varepsilon_p = 80$), ε_l is a dielectric constant of planar lipid bilayer ($\varepsilon_l = 2$), and ε_0 is vacuum permittivity ($\varepsilon_0 = 8.854 \times 10^{-12}$ Fm^{-1}). The thickness D of the planar lipid bilayer was considered to be37.5 nm and 43.2 nm for POPC and POPS, respectively, obtained from MD simulation as the head to head distance between two electron density profiles [33]. Due to lack of data considering D of POPC:POPS mixtures in scientific literature, we determined it according to MD simulation studies which deal with structure, dynamics and hydration dynamics of the mixed lipids [34,35]. Kastel et al. showed presence of domains of POPC and POPS lipids in planar lipid bilayers formed from POPC:POPS (4:1) mixtures [35]. The value 40.0 nm was assumed for the POPC:POPS mixture. Area of the planar lipid bilayer A is

$$\sum_{i=1}^{n} A_{pi} + \sum_{j=1}^{m} A_{lj} = A.$$

(9)

Therefore, Equation (8) can be rewritten as

$$C_{br} = \frac{\varepsilon_0}{D} \left(\varepsilon_l A + \sum_{i=1}^{n} A_{pi} (\varepsilon_p - \varepsilon_l) \right).$$

(10)

Change in planar lipid bilayer capacitance ΔC due to the pore formation is

$$\Delta C = C_{br} - C = -\frac{\varepsilon_0}{D} \sum_{i=1}^{n} A_{pi} \cdot (\varepsilon_p - \varepsilon_l),$$

(11)

if we assume that capacitance of planar lipid bilayer C at the beginning of the experiment is $C = \varepsilon_l \varepsilon_0 \frac{A}{D}$. To obtain the fraction of the planar lipid bilayer that is occupied by pores (A_{wat} / A), $\sum_{i=1}^{n} A_{pi}$ is calculated from ΔC and normalized on the whole area of planar lipid bilayer A.

We calculated ΔC from $\frac{d(C(t))}{dt}|_{t_{br}}$. It was multiplied by the planar lipid bilayer time constant t_c, which was calculated from experimental data in accordance with Equation (2).

3. Results

3.1. Experimental Results

The lipid composition c_{BLM}, measured by the chosen two methods, was not statistically different (Table 1). Moreover, also the reproducibility of both measuring methods were comparable. Our results, thus, confirm that these two measuring methods are equivalent.

Table 1. Specific capacitance of planar lipid bilayers (c_{BLM}) measured by discharge and capacitance to a period converting method. The values are given as mean $\pm$ standard deviation (number of measurements).

Lipid Mixtures	Discharge Method [25] c_{BLM} [μF/cm^2]	Capacitance to Period Converting Method [26] c_{BLM} [μF/cm^2]
POPC	0.51 ± 0.17 (80)	0.51 ± 0.16 (58) *
POPS	0.41 ± 0.14 (76)	0.41 ± 0.13 (34) *
POPC:POPS 1:1	0.31 ± 0.07 (60)	0.34 ± 0.17 (25) *

* $p > 0.05$.

Voltage breakdown for each lipid composition was measured by means of linearly rising electrical signal. Fast transmembrane voltage build-up was obtained by voltage-controlled experiments, while for a slow rise of transmembrane voltage current-controlled measurements were used. In Tables 2–4, U_{br} values are presented for fast and slow rise of transmembrane voltage, for all three lipid compositions, POPC, POPS, and POPC:POPS mixture, respectively. U_{br} values, measured using fast rise of transmembrane voltage, follow Equation (2) (Figure 2, left part), while U_{br} values measured using slow rise of transmembrane voltage exhibit high variability (Figure 2, right part). Therefore, we assumed that U_{br} did not depend on the slope of linearly rising current anymore and simply calculated the mean value of breakdown voltages U_{brI} for each lipid composition, regardless of the linearly rising current slope. The values of U_{brI} were 0.21 ± 0.04 V, 0.38 ± 0.11 V, and 0.46 ± 0.20 V for POPC, POPS, and POPC:POPS mixture, respectively. At these conditions, the U_{brI} values for planar lipid bilayers of different compositions are not statistically different. At slow rise of transmembrane voltage, the breakdown of planar lipid bilayer occurred in the range of seconds, which is 10^4 times slower in comparison with experiments using fast rise of transmembrane voltage (Figure 2). For all lipid compositions U_{brI} was lower than U_{brmin}, which was obtained by fitting Equation (2) to the results from experiments with fast rise of transmembrane voltage. The obtained values of U_{brmin} were 0.42 ± 0.01 V, 0.54 ± 0.02 V, and 0.38 ± 0.01 V for POPC, POPS, and POPC:POPS mixture, respectively, while values of t_c were 10.4 μs, 4.5 μs, 13.1 μs, presented in the same order.

Table 2. Results for planar lipid bilayers composed of POPC. In the upper part of the table, the results of voltage-controlled measurements with seven different slopes k_u are gathered while the results of current-controlled measurements with five different slopes k_i are presented in the lower part of the table. For each slope of linearly rising signal, the planar lipid breakdown voltage U_{br} and lifetime t_{br} are presented. Values are given as means $\pm$ standard deviations. $\frac{A_{wat}}{A}$ is a calculated fraction of the planar lipid bilayer that is occupied by pores. N is the number of experiments in each experimental group.

POPC					
	k_u (kV/s)	N	U_{br} (V)	t_{br} (μs)	$\frac{A_{wat}}{A}$ (%)
voltage-cont.	48.1	16	0.76 ± 0.05	16.76 ± 1.14	1.72
	21.6	17	0.67 ± 0.05	30.50 ± 2.60	0.95
	16.7	12	0.65 ± 0.06	39.38 ± 3.87	0.73
	11.5	12	0.59 ± 0.05	52.25 ± 4.65	0.55
	7.8	16	0.54 ± 0.04	70.74 ± 5.07	0.41
	5.5	16	0.54 ± 0.05	99.24 ± 9.93	0.29
	4.8	17	0.54 ± 0.04	112.92 ± 8.32	0.27
	k_i (μA/s)	N	U_{br} (V)	t_{br} (s)	$\frac{A_{wat}}{A}$ (%)
current-cont.	10	8	0.25 ± 0.14	0.64 ± 0.55	45.20×10^{-6}
	8	9	0.12 ± 0.16	1.22 ± 1.16	23.63×10^{-6}
	4	7	0.32 ± 0.16	1.63 ± 1.35	17.76×10^{-6}
	1	11	0.11 ± 0.11	3.30 ± 0.98	8.74×10^{-6}
	0.5	16	0.26 ± 0.13	5.97 ± 2.09	4.83×10^{-6}

Table 3. Results for planar lipid bilayers composed of POPS. In the upper part of the table, the results of voltage-controlled measurements with seven different slopes k_u are gathered while the results of current-controlled measurements with five different slopes k_i are presented in the lower part of the table. For each slope of linearly rising signal the planar lipid breakdown voltage U_{br} and lifetime t_{br} are presented. Values are given as means $\pm$ standard deviations. $\frac{A_{wat}}{A}$ is a calculated fraction of the planar lipid bilayer that is occupied by pores. N is the number of experiments in each experimental group.

	k_u (kV/s)	N	U_{br} (V)	t_{br} (µs)	$\frac{A_{wat}}{A}$ (%)
	POPS				
voltage-cont.	48.1	18	0.80 ± 0.04	17.59 ± 0.86	0.65
	21.6	14	0.72 ± 0.06	33.15 ± 2.76	0.35
	16.7	13	0.67 ± 0.05	41.24 ± 3.34	0.28
	11.5	12	0.66 ± 0.09	58.66 ± 7.71	0.20
	7.8	15	0.62 ± 0.08	81.05 ± 10.92	0.14
	5.5	13	0.59 ± 0.05	108.78 ± 8.72	0.11
	4.8	15	0.61 ± 0.04	129.24 ± 7.86	0.09
	k_i (µA/s)	N	U_{br} (V)	t_{br} (s)	$\frac{A_{wat}}{A}$ (%)
current-cont.	10	5	0.33 ± 0.04	0.61 ± 0.26	18.91×10^{-6}
	8	5	0.37 ± 0.04	0.98 ± 0.08	11.72×10^{-6}
	4	6	0.45 ± 0.14	2.37 ± 0.71	4.84×10^{-6}
	1	5	0.33 ± 0.04	6.99 ± 0.91	1.64×10^{-6}
	0.5	6	0.43 ± 0.14	17.98 ± 5.85	0.64×10^{-6}

Table 4. Results for planar lipid bilayers composed of POPC:POPS. In the upper part of the table, the results of voltage-controlled measurements with seven different slopes k_u are gathered while the results of current-controlled measurements with five different slopes k_i are presented in the lower part of the table. For each slope of linearly rising signal, the planar lipid breakdown voltage U_{br} and lifetime t_{br} are presented. Values are given as means $\pm$ standard deviations. $\frac{A_{wat}}{A}$ is a calculated fraction of the planar lipid bilayer that is occupied by pores. N is the number of experiments in each experimental group.

	k_u (kV/s)	N	U_{br} (V)	t_{br} (µs)	$\frac{A_{wat}}{A}$ (%)
	POPC:POPS 1:1				
voltage-cont.	48.1	7	0.72 ± 0.03	15.73 ± 0.73	1.49
	21.6	7	0.63 ± 0.05	28.90 ± 2.30	0.81
	16.7	6	0.59 ± 0.06	35.73 ± 3.70	0.66
	11.5	7	0.55 ± 0.04	48.49 ± 3.55	0.48
	7.8	11	0.53 ± 0.04	96.66 ± 4.60	0.24
	5.5	9	0.53 ± 0.05	96.68 ± 9.12	0.24
	4.8	13	0.48 ± 0.03	101.24 ± 7.39	0.23
	k_i (µA/s)	N	U_{br} (V)	t_{br} (s)	$\frac{A_{wat}}{A}$ (%)
current-cont.	10	3	0.27 ± 0.08	0.61 ± 0.19	15.92×10^{-6}
	8	5	0.52 ± 0.23	1.47 ± 0.51	38.47×10^{-6}
	4	5	0.63 ± 0.17	3.56 ± 0.94	6.59×10^{-6}
	1	6	0.43 ± 0.16	9.90 ± 3.70	2.37×10^{-6}
	0.5	5	0.47 ± 0.17	20.47 ± 8.92	1.15×10^{-6}

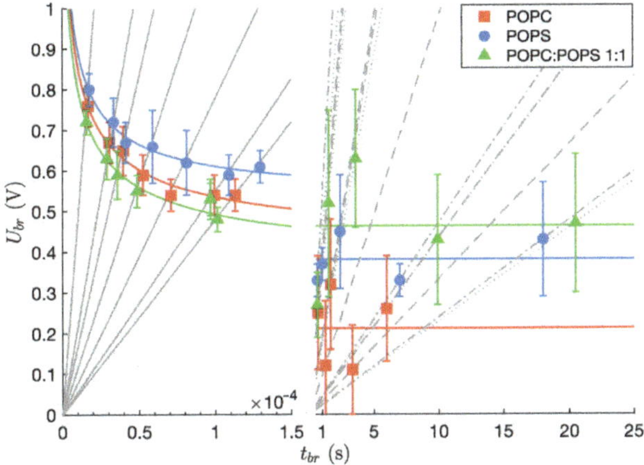

Figure 2. Breakdown voltages of specific lipid compositions measured during voltage-controlled (**left** side) and current-controlled (**right** side) experiments. The gray lines show different slopes (k_u—**left** side, k_i—**right** side) of linearly rising signal. Colored curves (**left** side) represent two-parameter strength-duration curves (Equation (2)) fitted to the data obtained by voltage-controlled measurements on POPS, POPC, and POPC:POPS mixture, respectively. Colored lines (**right** side) represent the average values of measured U_{brI} for POPC, POPS, and POPC:POPS mixture, respectively.

3.2. Modeling Results

The rate of the planar lipid bilayer capacitance change at t_{br} was calculated for all three lipid compositions. The results are presented in Figure 3 (left side). Planar lipid breakdowns that occur at slower transmembrane voltage build-up are accompanied by very small rates of the planar lipid bilayer capacitance change ($10^{-12} \leq \frac{dC}{dt} \leq 10^{-10}$ F/s). They do not follow any specific pattern. While the changes in the planar lipid bilayer capacitance for POPC planar lipid bilayer occur in the 0.11–0.32 V interval and for POPS planar lipid bilayer in 0.33–0.45 V interval, the changes in the planar lipid bilayer capacitance for POPC:POPS mixture are present also at higher voltages ($\leq$0.65 V). The rates of the planar lipid bilayer capacitance change, which accompany fast transmembrane voltage rise, are considerably greater ($10^{-7} \leq \frac{dC}{dt} \leq 10^{-5}$ F/s). As reflected in Figure 3 (right side), the rates of change of the planar lipid bilayer capacitance, in all specific compositions, are exponentially dependent on U_{br}: $\ln\left(\frac{dC}{dt}\right) = m \cdot U_{br} + n$. Referring to the kinetic model for membrane failure at fast loading rates, mentioned by Evans et al. [24], the parameter m is related to the pore radius r, in accordance with the equation $m = \frac{\pi r^2}{k_B T}$, where k_B is Boltzmann constant and T is an absolute temperature. Estimated pore radii at the breakdown moments at room temperature ($T = 298$ K), following fast rise of transmembrane voltage were 0.101 nm, 0.110 nm, and 0.106 nm, for planar lipid bilayers composed of POPC, POPS, and POPC:POPS, respectively. Parameter $\exp(n)$ can be considered as a spontaneous rate of the capacitance change or a spontaneous rate of water pores formation [24]. Its calculated value is $9.1 \cdot 10^{-9}$ F/s, $1.5 \cdot 10^{-9}$ F/s, and $4.6 \cdot 10^{-9}$ F/s for planar lipid bilayers composed of POPC, POPS, and POPC:POPS, respectively.

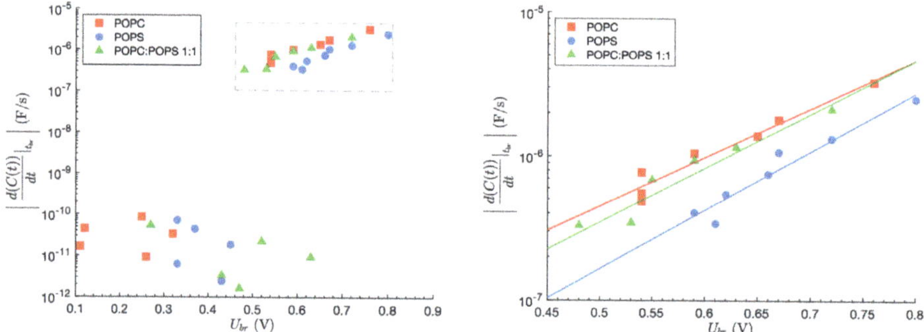

Figure 3. **Left** side of the figure presents the rate of the planar lipid bilayer capacitance change at t_{br} as a function of breakdown voltage U_{br}. Results regarding fast rise of transmembrane voltage are enlarged on the **right** side. Slope m of the linear relation between $\ln\left(\frac{dC}{dt}\right)$ and U_{br} is related to pore radius. See text for details.

In Figure 4, the fraction of the planar lipid bilayer area that is occupied by water pores is presented as a function of U_{br}. In the case of slow transmembrane voltage rise, which results in low values of U_{br}, the area of the planar lipid bilayer that is occupied by water pores is in the range of $10^{-6}\%$. The fraction increases to a couple of percent in case of fast transmembrane voltage rise, which results in higher values of U_{br} (see also data in Tables 2–4).

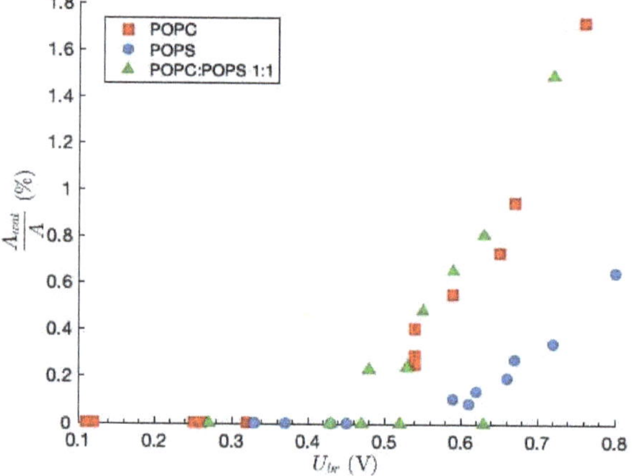

Figure 4. The fraction of the planar lipid bilayer area that is occupied by water pores as a function of U_{br}.

4. Discussion

In this study, planar lipid bilayers were exposed to linearly rising voltage and current signals of various slopes to examine U_{br}, the change in C at t_{br} and the fraction of the planar lipid bilayer occupied by water pores close to the membrane rupture. Three planar lipid bilayer compositions were used; non-charged zwitterionic POPC with mostly zero spontaneous curvature, negatively charged POPS with negative spontaneous curvature, and a mixture of both lipids at 1:1 ratio.

According to the fact that special attention should be given to C of each planar lipid bilayer, which is included in experimental studies [8], we measured C by two different measuring methods. Using a widely established discharge method and voltage to period conversion method, we proved that both methods give the same results and their reproducibility is comparable.

We found that within the experimental error, the specific capacitance of the POPC planar lipid bilayer was quite similar (Table 1) to that obtained for POPC bilayers reported previously ($c_{BLM} \approx 0.6\ \mu F/cm^2$) [36–38], although much lower values ($c_{BLM} = 0.28\ \mu F/cm^2$) have also been reported [39]. In addition, the specific capacitance measured for POPS planar lipid bilayers was slightly lower than values reported in the literature for lipids of similar type ($c_{BLM} = 0.624\ \mu F/cm^2$) [40]. The measured specific capacitance for planar lipid bilayers formed from a POPC:POPS mixture was considerably lower than previously reported values ($c_{BLM} = 0.6\ \mu F/cm^2$) [41]. For all three lipid compositions, we measured lower values of the specific capacitance. The most likely due to the use of a small diameter hole ($d = 117\ \mu m$), where the torus is not negligible and therefore the true planar lipid bilayer area is smaller than the geometric area associated with the hole diameter. Micelli et al. [42] studied the capacitance dependence on the hole radius and gave a simple relation for the width of the torus evaluation. If we consider $c_{BLM} = 0.6\ \mu F/cm^2$ as an appropriate value for the specific capacitance of POPC:POPS planar lipid bilayers, the width of the torus is 16.3 μm, which is reasonable. We think that the width of the torus can be dependent also on the planar lipid bilayer composition due to lipid molecules packing differences.

From U_{br} measurements presented in Figure 2, it is clear that U_{br} depends on dynamics of transmembrane voltage build-up. Application of slow slopes of linearly rising stimulus ($k_i \leq 10\ \mu A/s$) caused high variability of U_{br}. The mean values of U_{br} for specific lipid composition, calculated from all measurements obtained by slow slopes of linearly rising stimulus, were not statistically different. It seems that at slow slopes of linearly rising stimulus, the planar lipid bilayer rupture process, which occurs in the range of seconds, is stochastic to a great extent. At constant voltage applied on the planar lipid bilayer, fluctuations of conductance spikes were noticed [19] and associated with defects in planar lipid bilayer that appear and vanish spontaneously. Similarly, we noticed small voltage drops during constant or slowly linearly rising stimulus and reported about them in our previous studies [22,43,44]. Such conductive defects that appear and due to slowly rising stimulus are able also to vanish, locally discharge the planar lipid bilayer and can be the reason for the high variability of U_{br} in slow slopes of linearly rising stimulus conditions. MD simulations support the occurrence of a hydrophilic pore in planar lipid bilayers made from POPC [22] while POPS planar lipid bilayers exhibited only hydrophobic pores [44]. Although the appearance of a pore in POPS planar lipid bilayers was hydrophobic, ions were driven through the pore and discharged the planar lipid bilayer. Full recovery of the planar lipid bilayer in the case of hydrophilic pores is according to MD simulations tens of times longer as in the case of hydrophobic pores, whose lifetime is only a few nanoseconds. This is consistent with experimentally noticed voltage drops in POPS planar lipid bilayers being observed less frequently [44]. Similar observations were reported also on droplet interface planar lipid bilayers, where a population of pores of various sizes were visualized at constant voltage applied [45]. It was also reported that bilayer rupture takes place via a single electropore which grows while other pores shrink and close [45]. The lowest values of U_{br} were obtained for POPC planar lipid bilayers. In some cases of slowly increasing transmembrane voltage, planar lipid bilayers ruptured at transmembrane voltages close to 100 mV, which is in the range of the electric potential difference across a range of biological membranes [46]. Biological membranes, however, are not homogeneous single-component bilayers, but heterogeneous media with a complex organization of lipids and proteins. If we consider planar lipid bilayers of a POPC:POPS (1:1) mixture as a very simple model of a multicomponent bilayer, the experimental results showed that U_{br} is always higher than 200 mV, which can be an approximation of the upper value of the resting potential of biological membranes [47].

Application of fast slopes of linearly rising stimulus ($4.8\ kV/s \leq k_u \leq 48.1\ kV/s$) resulted in typical "strength-duration" behavior of U_{br}. The steeper the slope of the linearly rising stimulus, the higher the transmembrane voltage that is needed for planar lipid bilayer rupture. We fitted the strength–duration curve proposed by Dimitrov [30] to (t_{br}, U_{br}) points measured for all three lipid compositions. The chosen strength–duration curve considers the planar lipid bilayer as a thin viscoelastic film with fluctuating surfaces and was originally used to predict the critical voltage and critical time needed to breakdown the planar lipid bilayer at applied constant voltage. Planar lipid bilayers composed of POPS ruptured at higher voltages than planar lipid bilayers composed of POPC or the POPC:POPS mixture. According to the study on POPS/POPC (1:20) vesicles [48] and according to the study on solid-supported membranes of POPC:POPS (4:1) [35], irregularly shaped domains are formed in bilayers of POPC-POPS lipid mixtures. The formation of anionic lipid domains is mediated by the presence of cations in the environment; experiments and MD have shown that Ca^{2+} ions force the formation of larger PS domains [49] and it is also known that monovalent ions are able to induce domain formation [50]. However, charged lipids themselves also form segregated zones in membranes [51]. Moreover, the phase transition temperatures of PS lipids are always higher than those of its PC counterparts with the same acyl chains (e.g., that of POPC is -2 °C and that of POPS is $+14$ °C [52]), so incomplete lateral mixing of POPC and POPS is expected.

Measurements based on 2H nuclear magnetic resonance [48,53] have shown that the lipid hydrocarbon chains in pure PC and PS membranes have different order parameters at room temperatures. Namely, although both were in a lamellar liquid crystalline phase, PS was more ordered than PC. This difference is also reflected in differences in the lipid–lipid interaction at the lipid–water interface; namely, in the presence of an electric field, the incorporation of water into the hydrophobic region of the planar lipid bilayer is significantly enhanced in the disordered domains and almost unchanged in the ordered domains [54]. From molecular dynamics simulation studies, two regions in bilayers consisting of different domains are suggested for pore formation in the presence of an electric field: lipid regions with greater disorder [54,55] and thinned regions at domain boundaries [56]. According to Reigada et al. [54], pores are formed in lipid regions with greater disorder, and the smaller the disordered domains, the faster they are electroporated.. Our results are in line with these observations. Indeed, U_{brmin} were 0.42 ± 0.01 V and 0.54 ± 0.02 V for POPC and POPS planar lipid bilayers, respectively, implying that pores form more easily in more disordered POPC planar lipid bilayers. Moreover, for POPC planar lipid bilayers, the estimated spontaneous rate of water pore formation is six times higher than in the case of POPS planar lipid bilayers. U_{brmin} obtained for POPC:POPS mixture was even lower, 0.38 ± 0.01 V, which is consitent with the fact that smaller disordered domains are more susceptible for pore formation. On the other hand, the formation of pores could also be expected at the domain boundaries, where near the phase transition of single-component DPPC planar lipid bilayer, pore formation induced by an external electric field was obtained in thinned regions at the liquid and gel-domain interfaces [56]. Although our experiments were performed far from the transition temperature (at least 10 °C), thinner regions with reduced stability [57] can also be attributed to boundaries between disordered POPC and more ordered POPS lipid domains.

The rate of capacitance change of the planar lipid bilayer can be related to the rate of water pores formation [14,16,18]. Ruptures of planar lipid bilayer that occur as a consequence of a slow build-up of transmembrane voltage are accompanied by very small rates of the capacitance change, which do not exhibit any specific pattern. However, the rates of capacitance change which appear at fast build-up of transmembrane voltage are exponentially dependent on U_{br}. Similarly, Evans et al. [24] reported that the size and frequency of initial defect formation of fluid membrane vesicles depend on the loading rate of micro-pipette suction. Estimated radii of pores that govern ruptures of a planar lipid bilayer are 0.101 nm, 0.110 nm, and 0.106 nm for planar lipid bilayers composed of POPC, POPS, and POPC:POPS, respectively. The values are in the range of radii for hydrophobic pores predicted by Akimov's model [15]. The model also predicts a slightly larger radius of hydrophobic pores for planar lipid bilayers composed of lipids with negative spontaneous curvature while compared with planar lipid bilayers composed of lipids with zero spontaneous curvature, which is in agreement with our experiments and modeling.

Estimation of the area fraction that is occupied with water pores at the moment of planar lipid bilayer rupture reveals that the number of water pores is considerably lower in the case of slow build-up of transmembrane voltage. It is probably due to the fact that during slow rise of the transmembrane voltage, conducting pores are already present in planar lipid bilayer [19,45], which also changes the electrical resistance of planar lipid bilayer. It should be noted that this change was not taken into account in our calculations. Aqueous fractional area in the planar lipid bilayer due to an exponential pulse, bipolar square pulse, or a square pulse was theoretically evaluated by Freeman et al. in 1994 [14]. According to their calculations, only 0.1% of the membrane is occupied by pores, which is in good agreement with our results obtained for a fast build-up of transmembrane voltage. Moreover, recently, Anosov et al. [18] presented a computational model which predicts the number of hydrophobic pores in planar lipid bilayer using the variance of the membrane current. Their calculations showed that the area of hydrophobic pores at the phase transition (absence of a stimulus) does not exceed 4.5% of the planar lipid bilayer area, which is also in agreement with our observations.

The area fraction that is occupied with water pores is similar in the case of planar lipid bilayers formed of POPC and POPC:POPS mixture, while POPS planar lipid bilayers exhibited lower values. Lipid molecules in POPS planar lipid bilayers are more ordered than in POPC planar lipid bilayers [58] and according to MD simulations in POPS planar lipid bilayers mostly hydrophobic pores are formed where lifetime is only a few nanoseconds [44], allowing quick planar lipid bilayer discharge. As a result that pores are harder to form in ordered planar lipid bilayer, also our results show a six times lower spontaneous rate of water pore formation in comparison with POPC lipids. Therefore, it is possible that the area fraction that is occupied with water pores at the moment of planar lipid bilayer rupture in POPS planar lipid bilayers is indeed noticeably smaller in comparison with POPC and POPC:POPS planar lipid bilayers. The area fraction occupied with water pores at the moment of the planar lipid bilayers rupture is similar in the case of POPC and POPC:POPS planar lipid bilayers. Although it seems that in the planar lipid bilayers made of POPC:POPS mixture, pores are predominantly formed in more disordered POPC domains, additional experimental and theoretical studies are needed to confirm this guess.

Estimation of the radius of a water pore that governs planar lipid bilayer rupture and the fraction of planar lipid bilayer area that is occupied by water pores at the moment of rupture is based on prediction that small water pores are present in the planar lipid bilayer. Therefore, the difference in the dielectric permittivity of planar lipid bilayer ($\varepsilon = 2$) and water ($\varepsilon = 80$) is the only reason for the change in capacitance. This assumption is a considerable simplification of all phenomena that are involved in water pore creation and existence. Close to the charged surface, for example, strong orientational water dipole ordering may result in a local decrease of permittivity [59,60]. The relative permittivity of the electrolyte solution decreases with the increasing magnitude of the electric field strength. In narrow tubes, like in water pores, the water dipoles are partially oriented also close to the axis of the tube, which decreases the relative water permittivity [61]. To obtain more reliable results, at least orientational ordering of water molecules in pores and close to the membrane should be taken into account.

5. Conclusions

Theories of pore formation and molecular dynamics simulations suggest recurrent hydrophobic pores or water pores whose occurrence is strongly dependent on the electric field. Experimental information about them is very rare. In our study, planar lipid bilayers of POPC, POPS, and a POPC:POPS (1:1) mixture were exposed to different slopes of linearly increasing voltage or current signal. Such experiments allowed to determine the breakdown voltage of planar lipid bilayers, and taking into account that the membrane can be represented by an RC circuit, the fraction of the planar lipid bilayer occupied by water pores could be calculated. Our results confirm the results of pore formation theories and molecular dynamics models. The fraction of area occupied by water pores at the moment of planar lipid bilayer rupture is considerably lower when the transmembrane voltage is built up slower than in the case of a fast build-up of transmembrane voltage. In more disordered planar lipid bilayers, water pores are easily formed, and the radii of the water pores are smaller in this case than in more ordered planar lipid bilayers. In the case of lipid mixtures, water pores are formed even more easily, probably due to smaller disordered domains in the planar lipid bilayer or/and weaker regions at the domain boundaries.

Author Contributions: Conceptualization, P.K. and A.M.L.; methodology, P.K. and A.M.L.; validation, P.K. and A.M.L.; formal analysis, P.K. and A.M.L.; investigation, P.K. and A.M.L.; data curation, P.K. and A.M.L.; writing—original draft preparation, P.K., A.M.L., M.K., and D.M.; visualization, P.K.; supervision, P.K. and A.M.L.; funding acquisition, D.M. and M.K. All authors have read and agreed to the published version of the manuscript.

Funding: This research was supported by Slovenian Research Agency (ARRS) research core funding No. P2-0249. The part of experimental work was performed in the scope of Polish-Slovenian Scientific Cooperation: Electroporation of cells in a current-controlled mode funded by Ministry of Science and Higher Education (MNiSW), Poland and Slovenian Research Agency (ARRS), Slovenia.

Institutional Review Board Statement: Not applicable.

Informed Consent Statement: Not applicable.

Data Availability Statement: All data presented are available in this article.

Conflicts of Interest: The authors declare no conflict of interest.

Abbreviations

The following abbreviations are used in this manuscript:

POPC	1-pamitoyl 2-oleoyl phosphatidylcholine
POPS	1-pamitoyl 2-oleoyl phosphatidylserine
MD	Molecular dynamics simulation
GUV	Giant unilamellar vesicles

References

1. Siontorou, C.; Nikoleli, G.P.; Nikolelis, D.; Karapetis, S. Artificial Lipid Membranes: Past, Present, and Future. *Membranes* **2017**, *7*, 38. [CrossRef] [PubMed]
2. Brosseau, C.; Sabri, E. Resistor–capacitor modeling of the cell membrane: A multiphysics analysis. *J. Appl. Phys.* **2021**, *129*, 011101. [CrossRef]
3. Naumowicz, M.; Figaszewski, Z.A. Pore Formation in Lipid Bilayer Membranes made of Phosphatidylcholine and Cholesterol Followed by Means of Constant Current. *Cell Biochem. Biophys.* **2013**, *66*, 109–119. [CrossRef] [PubMed]
4. Kramar, P.; Miklavčič, D.; Kotulska, M.; Lebar, A. *Voltage- and Current-Clamp Methods for Determination of Planar Lipid Bilayer Properties*; Elsevier Inc.: Amsterdam, The Netherlands, 2010; Volume 11. [CrossRef]
5. Naumowicz, M.; Kotynska, J.; Petelska, A.; Figaszewski, Z. Impedance analysis of phosphatidylcholine membranes modified with valinomycin. *Eur. Biophys. J.* **2006**, *35*, 239–246. [CrossRef]
6. Shlyonsky, V.; Dupuis, F.; Gall, D. The OpenPicoAmp: An Open-Source Planar Lipid Bilayer Amplifier for Hands-On Learning of Neuroscience. *PLoS ONE* **2014**, *9*, e108097. [CrossRef]
7. Mosgaard, L.D.; Zecchi, K.A.; Heimburg, T.; Budvytyte, R. The effect of the nonlinearity of the response of lipid membranes to voltage perturbations on the interpretation of their electrical properties. A new theoretical description. *Membranes* **2015**, *5*, 495–512. [CrossRef]
8. Pavlin, M.; Kotnik, T.; Miklavčič, D.; Kramar, P.; Maček Lebar, A. Chapter Seven Electroporation of Planar Lipid Bilayers and Membranes. *Adv. Planar Lipid Bilayers Liposomes* **2008**, *6*, 165–226. [CrossRef]
9. Kotnik, T.; Rems, L.; Tarek, M.; Miklavčič, D. Membrane Electroporation and Electropermeabilization: Mechanisms and Models. *Annu. Rev. Biophys.* **2019**, *48*, 63–91. [CrossRef]
10. Hu, Y.; Sinha, S.K.; Patel, S. Investigating Hydrophilic Pores in Model Lipid Bilayers Using Molecular Simulations: Correlating Bilayer Properties with Pore-Formation Thermodynamics. *Langmuir* **2015**, *31*, 6615–6631. [CrossRef]
11. Alvarez, O.; Latorre, R. Voltage-dependent capacitance in lipid bilayers made from monolayers. *Biophys. J.* **1978**, *21*, 1–17. [CrossRef]
12. Heimburg, T. The capacitance and electromechanical coupling of lipid membranes close to transitions: The effect of electrostriction. *Biophys. J.* **2012**, *103*, 918–929. [CrossRef]

13. Sabri, E.; Lasquellec, S.; Brosseau, C. Electromechanical modeling of the transmembrane potential-dependent cell membrane capacitance. *Appl. Phys. Lett.* **2020**, *117*, 043701. [CrossRef]
14. Freeman, S.A.; Wang, M.A.; Weaver, J.C. Theory of electroporation of planar bilayer membranes: Predictions of the aqueous area, change in capacitance, and pore-pore separation. *Biophys. J.* **1994**, *67*, 42–56. [CrossRef]
15. Akimov, S.A.; Volynsky, P.E.; Galimzyanov, T.R.; Kuzmin, P.I.; Pavlov, K.V.; Batishchev, O.V. Pore formation in lipid membrane II: Energy landscape under external stress. *Sci. Rep.* **2017**, *7*, 12509. [CrossRef]
16. Abidor, I.G.; Arakelyan, V.B.; Chernomordik, L.V.; Chizmadzhev, Y.A.; Pastushenko, V.F.; Tarasevich, M.P. Electric breakdown of bilayer lipid membranes. I. The main experimental facts and their qualitative discussion. *J. Electroanal. Chem.* **1979**, *104*, 37–52. [CrossRef]
17. Tieleman, D.P.; Leontiadou, H.; Mark, A.E.; Marrink, S.J. Simulation of Pore Formation in Lipid Bilayers by Mechanical Stress and Electric Fields. *J. Am. Chem. Soc.* **2003**, *125*, 6382–6383. [CrossRef]
18. Anosov, A.A.; Smirnova, E.Y.; Sharakshane, A.A.; Nikolayeva, E.A.; Zhdankina, Y.S. Increase in the current variance in bilayer lipid membranes near phase transition as a result of the occurrence of hydrophobic defects. *Biochim. Biophys. Acta Biomembr.* **2020**, *1862*, 183147. [CrossRef]
19. Melikov, K.C.; Frolov, V.A.; Shcherbakov, A.; Samsonov, A.V.; Chizmadzhev, Y.A.; Chernomordik, L.V. Voltage-Induced Nonconductive Pre-Pores and Metastable Single Pores in Unmodified Planar Lipid Bilayer. *Biophys. J.* **2001**, *80*, 1829–1836. [CrossRef]
20. Kalinowski, S.; Ibron, G.; Bryl, K.; Figaszewski, Z. Chronopotentiometric studies of electroporation of bilayer lipid membranes. *Biochim. Biophys. Acta Biomembr.* **1998**, *1369*, 204–212. [CrossRef]
21. Kotulska, M. Natural fluctuations of an electropore show fractional Lévy stable motion. *Biophys. J.* **2007**, *92*, 2412–2421. [CrossRef]
22. Kramar, P.; Delemotte, L.; Lebar, A.M.; Kotulska, M.; Tarek, M.; Miklavčič, D. Molecular-level characterization of lipid membrane electroporation using linearly rising current. *J. Membr. Biol.* **2012**, *245*, 651–659. [CrossRef] [PubMed]
23. Akimov, S.A.; Volynsky, P.E.; Galimzyanov, T.R.; Kuzmin, P.I.; Pavlov, K.V.; Batishchev, O.V. Pore formation in lipid membrane I: Continuous reversible trajectory from intact bilayer through hydrophobic defect to transversal pore. *Sci. Rep.* **2017**, *7*, 1–20. [CrossRef] [PubMed]
24. Evans, E.; Heinrich, V.; Ludwig, F.; Rawicz, W. Dynamic tension spectroscopy and strength of biomembranes. *Biophys. J.* **2003**, *85*, 2342–2350. [CrossRef]
25. Kramar, P.; Miklavcic, D.; Lebar, A.M. Determination of the lipid bilayer breakdown voltage by means of linear rising signal. *Bioelectrochemistry* **2007**, *70*, 23–27. [CrossRef] [PubMed]
26. Kalinowski, S.; Figaszewski, Z. A four-electrode system for measurement of bilayer lipid membrane capacitance. *Meas. Sci. Technol.* **1995**, *6*, 1043–1049. [CrossRef]
27. Kalinowski, S.; Figaszewski, Z. A four-electrode potentiostat-galvanostat for studies of bilayer lipid membranes. *Meas. Sci. Technol.* **1995**, *6*, 1050–1055. [CrossRef]
28. Montal, M.; Mueller, P. Formation of Bimolecular Membranes from Lipid Monolayers and a Study of Their Electrical Properties. *Proc. Natl. Acad. Sci. USA* **1972**, *69*, 3561–3566. [CrossRef]
29. Kramar, P.; Miklavčič, D.; Lebar, A.M. A system for the determination of planar lipid bilayer breakdown voltage and its applications. *IEEE Trans. Nanobiosci.* **2009**, *8*, 132–138. [CrossRef]
30. Dimitrov, D.S. Electric field-induced breakdown of lipid bilayers and cell membranes: A thin viscoelastic film model. *J. Membr. Biol.* **1984**, *78*, 53–60. [CrossRef]
31. Sabotin, I.; Lebar, A.; Miklavčič, D.; Kramar, P. Measurement protocol for planar lipid bilayer viscoelastic properties. *IEEE Trans. Dielectr. Electr. Insul.* **2009**, *16*, 1236–1242. [CrossRef]
32. Levine, Z.A.; Vernier, P.T. Life cycle of an electropore: Field-dependent and field-independent steps in pore creation and annihilation. *J. Membr. Biol.* **2010**, *236*, 27–36. [CrossRef]
33. Shahane, G.; Ding, W.; Palaiokostas, M.; Orsi, M. Physical properties of model biological lipid bilayers: Insights from all-atom molecular dynamics simulations. *J. Mol. Model.* **2019**, *25*, 1–13. [CrossRef]
34. Jurkiewicz, P.; Cwiklik, L.; Vojtíšková, A.; Jungwirth, P.; Hof, M. Structure, dynamics, and hydration of POPC/POPS bilayers suspended in NaCl, KCl, and CsCl solutions. *Biochim. Biophys. Acta Biomembr.* **2012**, *1818*, 609–616. [CrossRef]
35. Kastl, K.; Menke, M.; Lüthgens, E.; Faiß, S.; Gerke, V.; Janshoff, A.; Steinem, C. Partially reversible adsorption of annexin A1 on POPC/POPS bilayers investigated by QCM measurements, SFM, and DMC simulations. *ChemBioChem* **2006**, *7*, 106–115. [CrossRef]
36. Troiano, G.C.; Tung, L.; Sharma, V.; Stebe, K.J. The reduction in electroporation voltages by the addition of a surfactant to planar lipid bilayers. *Biophys. J.* **1998**, *75*, 880–888. [CrossRef]
37. Naumowicz, M.; Petelska, A.D.; Figaszewski, Z.A. Physicochemical analysis of phosphatidylcholine-ceramide system in bilayer lipid membranes. *Acta Biochim. Pol.* **2008**, *55*, 721–730._3033. [CrossRef]
38. Cannon, B.; Hermansson, M.; Györke, S.; Somerharju, P.; Virtanen, J.A.; Cheng, K.H. Regulation of calcium channel activity by lipid domain formation in planar lipid bilayers. *Biophys. J.* **2003**, *85*, 933–942. [CrossRef]
39. Meleleo, D. Study of Resveratrol's Interaction with Planar Lipid Models: Insights into Its Location in Lipid Bilayers. *Membranes* **2021**, *11*, 132. [CrossRef]

40. Naumowicz, M.; Figaszewski, Z.A. Impedance Spectroscopic Investigation of the Bilayer Lipid Membranes Formed from the Phosphatidylserine–Ceramide Mixture. *J. Membr. Biol.* **2009**, *227*, 67–75. [CrossRef]

41. Naumowicz, M.; Figaszewski, Z.A. The effect of pH on the electrical capacitance of phosphatidylcholine-phosphatidylserine system in bilayer lipid membrane. *J. Membr. Biol.* **2014**, *247*, 361–369. [CrossRef]

42. Micelli, S.; Gallucci, E.; Meleleo, D.; Stipani, V.; Picciarelli, V. Mitochondrial porin incorporation into black lipid membranes: Ionic and gating contribution to the total current. *Bioelectrochemistry* **2002**, *57*, 97–106. [CrossRef]

43. Kotulska, M.; Basalyga, J.; Derylo, M.B.; Sadowski, P. Metastable pores at the onset of constant-current electroporation. *J. Membr. Biol.* **2010**, *236*, 37–41. [CrossRef]

44. Dehez, F.; Delemotte, L.; Kramar, P.; Miklavčič, D.; Tarek, M. Evidence of conducting hydrophobic nanopores across membranes in response to an electric field. *J. Phys. Chem. C* **2014**, *118*, 6752–6757. [CrossRef]

45. Sengel, J.T.; Wallace, M.I. Imaging the dynamics of individual electropores. *Proc. Natl. Acad. Sci. USA* **2016**, *113*, 5281–5286. [CrossRef]

46. What Is the Electric Potential Difference Across Membranes? Available online: http://book.bionumbers.org/what-is-the-electric-potential-difference-across-membranes/ (accessed on 27 March 2021)

47. Kinoshita, N.; Unemoto, T.; Kobayashi, H. Proton motive force is not obligatory for growth of Escherichia coli. *J. Bacteriol.* **1984**, *160*, 1074–1077. [CrossRef]

48. Justice, M.J.; Petrusca, D.N.; Rogozea, A.L.; Williams, J.A.; Schweitzer, K.S.; Petrache, I.; Wassall, S.R.; Petrache, H.I. Effects of Lipid Interactions on Model Vesicle Engulfment by Alveolar Macrophages. *Biophys. J.* **2014**, *106*, 598–609. [CrossRef]

49. Sahoo, A.; Matysiak, S. Microscopic Picture of Calcium-Assisted Lipid Demixing and Membrane Remodeling Using Multiscale Simulations. *J. Phys. Chem. B* **2020**, *124*, 7327–7335. [CrossRef] [PubMed]

50. Ganesan, S.J.; Xu, H.; Matysiak, S. Influence of Monovalent Cation Size on Nanodomain Formation in Anionic–Zwitterionic Mixed Bilayers. *J. Phys. Chem. B* **2017**, *121*, 787–799. [CrossRef] [PubMed]

51. Bobone, S.; Hilsch, M.; Storm, J.; Dunsing, V.; Herrmann, A.; Chiantia, S. Phosphatidylserine Lateral Organization Influences the Interaction of Influenza Virus Matrix Protein 1 with Lipid Membranes. *J. Virol.* **2017**, *91*, 1–15. [CrossRef] [PubMed]

52. Phase Transition Temperatures for Glycerophospholipids. Available online: https://avantilipids.com/tech-support/physical-properties/phase-transition-temps (accessed on 27 March 2021).

53. Huster, D.; Arnold, K.; Gawrisch, K. Influence of docosahexaenoic acid and cholesterol on lateral lipid organization in phospholipid mixtures. *Biochemistry* **1998**, *37*, 17299–17308. [CrossRef]

54. Reigada, R. Electroporation of heterogeneous lipid membranes. *Biochim. Biophys. Acta Biomembr.* **2014**, *1838*, 814–821. [CrossRef]

55. López Martí, J.M.; English, N.J.; Del Pópolo, M.G. Elucidating mysteries of phase-segregated membranes: Mobile-lipid recruitment facilitates pores' passage to the fluid phase. *Phys. Chem. Chem. Phys.* **2018**, *20*, 19234–19239. [CrossRef]

56. Kirsch, S.A.; Böckmann, R.A. Coupling of Membrane Nanodomain Formation and Enhanced Electroporation near Phase Transition. *Biophys. J.* **2019**, *116*, 2131–2148. [CrossRef]

57. Cordeiro, R.M. Molecular Structure and Permeability at the Interface between Phase-Separated Membrane Domains. *J. Phys. Chem. B* **2018**, *122*, 6954–6965. [CrossRef]

58. Engberg, O.; Yasuda, T.; Hautala, V.; Matsumori, N.; Nyholm, T.K.; Murata, M.; Slotte, J.P. Lipid Interactions and Organization in Complex Bilayer Membranes. *Biophys. J.* **2016**, *110*, 1563–1573. [CrossRef]

59. Lebar, A.M.; Velikonja, A.; Kramar, P.; Iglič, A. Internal configuration and electric potential in planar negatively charged lipid head group region in contact with ionic solution. *Bioelectrochemistry* **2016**, *111*, 49–56. [CrossRef]

60. Iglič, A.; Gongadze, E.; Kralj-Iglič, V. Differential capacitance of electric double layer – Influence of asymmetric size of ions, thickness of stern layer and orientational ordering of water dipoles. *Acta Chim. Slov.* **2019**, *66*, 534–541. [CrossRef]

61. Drab, M.; Gongadze, E.; Kralj-Iglič, V.; Iglič, A. Electric double layer and orientational ordering of water dipoles in narrow channels within a modified Langevin Poisson-Boltzmann model. *Entropy* **2020**, *22*, 1054. [CrossRef]

Article

Dielectric Properties of Phosphatidylcholine Membranes and the Effect of Sugars

Victoria Vitkova [1,*], Vesela Yordanova [2], Galya Staneva [2], Ognyan Petkov [1], Angelina Stoyanova-Ivanova [1], Krassimira Antonova [1] and Georgi Popkirov [3]

[1] Georgi Nadjakov Institute of Solid State Physics, Bulgarian Academy of Sciences, 72 Tsarigradsko Chaussee, Blvd., 1784 Sofia, Bulgaria; ogikrpetkov@gmail.com (O.P.); angelina@issp.bas.bg (A.S.-I.); krasa@issp.bas.bg (K.A.)

[2] Institute of Biophysics and Biomedical Engineering, Bulgarian Academy of Sciences, Acad. G. Bonchev Str., Bl. 21, 1113 Sofia, Bulgaria; v.v.yordanova.bul@abv.bg (V.Y.); gstaneva@obzor.bio21.bas.bg (G.S.)

[3] Central Laboratory of Solar Energy and New Energy Sources, Bulgarian Academy of Sciences, 72 Tsarigradsko Chaussee, Blvd., 1784 Sofia, Bulgaria; popkirov@yahoo.com

* Correspondence: victoria@issp.bas.bg

Citation: Vitkova, V.; Yordanova, V.; Staneva, G.; Petkov, O.; Stoyanova-Ivanova, A.; Antonova, K.; Popkirov, G. Dielectric Properties of Phosphatidylcholine Membranes and the Effect of Sugars. *Membranes* **2021**, *11*, 847. https://doi.org/10.3390/membranes11110847

Academic Editor: Monika Naumowicz

Received: 30 September 2021
Accepted: 27 October 2021
Published: 30 October 2021

Abstract: Simple carbohydrates are associated with the enhanced risk of cardiovascular disease and adverse changes in lipoproteins in the organism. Conversely, sugars are known to exert a stabilizing effect on biological membranes, and this effect is widely exploited in medicine and industry for cryopreservation of tissues and materials. In view of elucidating molecular mechanisms involved in the interaction of mono- and disaccharides with biomimetic lipid systems, we study the alteration of dielectric properties, the degree of hydration, and the rotational order parameter and dipole potential of lipid bilayers in the presence of sugars. Frequency-dependent deformation of cell-size unilamellar lipid vesicles in alternating electric fields and fast Fourier transform electrochemical impedance spectroscopy are applied to measure the specific capacitance of phosphatidylcholine lipid bilayers in sucrose, glucose and fructose aqueous solutions. Alteration of membrane specific capacitance is reported in sucrose solutions, while preservation of membrane dielectric properties is established in the presence of glucose and fructose. We address the effect of sugars on the hydration and the rotational order parameter for 1-palmitoyl-2-oleoyl-*sn*-glycero-3- phosphocholine (POPC) and 1-stearoyl-2-oleoyl-*sn*-glycero-3- phosphocholine (SOPC). An increased degree of lipid packing is reported in sucrose solutions. The obtained results provide evidence that some small carbohydrates are able to change membrane dielectric properties, structure, and order related to membrane homeostasis. The reported data are also relevant to future developments based on the response of lipid bilayers to external physical stimuli such as electric fields and temperature changes.

Keywords: lipid bilayers; sucrose; capacitance; relative permittivity; dipole potential; membrane structure

1. Introduction

The biological significance of carbohydrate molecules and developments of their biomedical and industrial applications have stimulated intensive research toward gaining knowledge of the alteration of membrane physicochemical properties by the presence of sugars. The protection of higher plant cells from the consequences of prolonged dehydration with the help of oligo- and polysaccharides represents a prominent example of the stabilizing effect exerted by sugars on biological membranes [1–3]. Simple carbohydrates represent one of the major protecting excipients in preservation of cells during freeze-drying [4]. The cryoprotective efficiency of sugars is related to the molecular mechanisms of their interaction with lipid membranes. It has been established that even under conditions of tight lipid packing, polysaccharides are able to sustain the liquid-crystalline lamellar phase of phospholipids by penetrating between the lipid headgroups [5], which is coherent with their strong drought-protective role in cellular membranes [1]. Simple carbohydrates (mono- and disaccharides) are also found to exert a protective role against abnormal temperatures and dehydration in biomembranes and synthetic lipid bilayers [6]. Some disaccharides such as trehalose and sucrose are identified as efficient cryoprotectants counteracting the water deficiency via replacement of water molecules at the membrane surface [1,7,8]. Large decrease of the lateral phospholipid mobility by sucrose has been evidenced by fluorescence correlation spectroscopy measurements and molecular dynamic simulations [7]. The strong influence of simple carbohydrates and oligo- and polysaccharides on the tension of lipid monolayers has also been reported [8]. Data acquired for mechanical properties of bilayer lipid stacks [9,10] and free-standing lipid membranes [11–13] in the presence of mono- and disaccharides outline a more complex picture on sugar–membrane interactions.

The detailed characterization of the electrical properties of biological membranes contributes to the understanding of the governing mechanisms involved in cellular processes such as transmission of electrical impulses or synaptic activity. Research in this direction is relevant to numerous electroporation-based applications in drug delivery, electrochemotherapy, cell–cell hybridization, food processing, etc. [14]. The lipid bimolecular matrix represents the main structural entity of cell membranes. Since it is impermeable to ions, the lipid bilayer behaves as a capacitor, strongly influencing the electric field distribution in the cell. The quantification of the lipid bilayer capacitance and the estimation of its dependence on external factors allows for evaluating the charging time of membranes and membrane–field interactions, including transmembrane potential difference [15]. The alteration of the bilayer electrical capacitance in sucrose-containing solutions reported recently [16] emerged various questions regarding the effect on membrane electrical properties of dissolving simple carbohydrates in the aqueous surroundings.

Considering the close relation of the specific capacitance to membrane composition, its structure and physical state [17], we probe here the susceptibility of this important electrical parameter of lipid bilayers to the presence of sugar molecules in the aqueous surroundings. The effect of glucose, fructose and sucrose on the degree of hydration and rotational order parameter of lipid molecules is measured with regard to expected alterations of membrane dielectric permittivity [16,18–20]. The specific capacitance of free-standing membranes is measured by frequency-dependent electrodeformation of giant unilamellar lipid vesicles (GUVs) and Fast-Fourier transform electrochemical impedance spectrometry (FFT-EIS) of planar bilayer lipid membranes (BLMs) for acquisition of the impedance spectrum (1 Hz–50 kHz) in a couple of seconds. The simplest physical model of biological membranes is the lipid bilayer [21]. Characterized by diameters in the same range as the typical cell sizes (5–100 μm), GUVs are considered as a basic physical model of biomembranes allowing for investigation of membrane-related phenomena at the scale of single vesicles [22,23]. The elaborated experimental protocols for preparation of GUVs offer a good control of membrane composition and the physicochemical parameters of the aqueous environment. Membrane dipole potential, the degree of hydration and rotational order parameter of lipid molecules are studied in sugar-containing aqueous solutions by fluorescence spectroscopy of large unilamellar vesicles (LUVs), which are characterized by orders of magnitude smaller diameters (~100 nm) compared to GUVs. In addition to the application for modeling biomembranes, lipid vesicles are also recognized for their usage as drug and gene carriers [24], which foregrounds the investigation of their membrane physical properties and stability.

Here, we study the alteration of dielectric properties, the degree of hydration, the rotational order parameter and dipole potential of lipid bilayers in the presence of simple carbohydrates. For the first time, the effect of glucose and fructose is probed on the impedance characteristics of biomimetic lipid membranes. The reported results provide knowledge of sugar–membrane interactions in relation to the bilayer molecular organization, electrical capacitance, resistance and dielectric permittivity. The acquired new data about the alteration of the dielectric properties and molecular organization of lipid bilayers in sugar solutions are expected to facilitate the selection of the best carbohydrate to meet requirements of applications related to the response of lipid bilayers to external electric fields.

2. Materials and Methods

2.1. Materials

Synthetic monounsaturated phospholipids 1-palmitoyl-2-oleoyl-*sn*-glycero- 3-phosphocholine (16:0-18:1 PC, POPC) and 1-stearoyl-2-oleoyl-*sn*-glycero- 3-phosphocholine (18:0-18:1 PC, SOPC) purchased from Avanti Polar Lipids Inc. (Alabaster, AL, USA) in powder, are used to produce model bilayer systems as described further. Molecular probes for fluorescence spectroscopy measurements, namely 6-dodecanoyl-N, N-dimethyl-2-naphthylamine (Laurdan), 1,6-diphenyl-1,3,5-hexatriene (DPH) and 4-(2-[6-(Dioctylamino)-2-naphthalenyl] ethenyl)-1-(3-sulfopropyl)pyridinium inner salt (di-8-ANEPPS) are purchased from Sigma-Aldrich (Darmstadt, Germany). The same provider supplied chloroform and methanol for lipid solutions, sodium chloride for aqueous solutions as well as D-(+)-Glucose ($C_6H_{12}O_6$, BioXtra, $\geq$99.5%, GC), D-($-$)-Fructose ($C_6H_{12}O_6$, BioUltra, $\geq$99.0%, HPLC) and sucrose ($C_{12}H_{22}O_{11}$, BioXtra, $\geq$99.5%, GC). Pentane $CH_3(CH_2)_3CH_3$ and hexane $CH_3(CH_2)_4CH_3$ of HPLC grade are purchased from Honeywell, Riedel-de Haën (Seelze, Germany). Polydimethylsiloxane (PDMS) is provided by Dow Corning (Midland, MI, USA). All substances used for the preparation of bilayer membranes and vesicle suspensions are utilized as purchased. Aqueous solutions are prepared with bidistilled water from a quartz distiller.

2.2. Methods

2.2.1. Preparation of Giant Unilamellar Vesicles

Electroformation method is applied to produce giant unilamellar vesicles (GUVs) for measurements of membrane electrical capacitance. The electroformation cell consists of two indium tin oxide (ITO)-coated glass plates, serving as electrodes, which are separated by a PDMS spacer [25]. A small amount (~50 µg) of POPC or SOPC with lipid concentration of 1 g/L in chloroform-methanol solvent (9:1 volume parts) is uniformly spread on the ITO-coated side of each glass plate. After the complete evaporation of the organic solvents achieved under vacuum, the electroformation chamber is filled with 1 mmol/L NaCl aqueous solution, which is sugar-free (control) or contains up to 300 mmol/L of glucose, fructose or sucrose. AC electric field with 10 Hz frequency and peak-to-peak voltage amplitude successively increased to 4 V is applied to the chamber. A high yield of quasispherical unilamellar vesicles with diameters of dozens of micrometers is obtained in several (~3) hours. The conductivities of the aqueous solutions are measured with Cyberscan PC510 (Eutech Instruments, Singapore). Prior to electrodeformation measurements of GUVs, we add 0.1 mmol/L of sodium chloride to the suspension in order to increase the conductivity of the external solution in accordance with the requirements of the GUV electrodeformation method (see Section 2.2.5).

2.2.2. Preparation of Large Unilamellar Vesicles

Fluorescence spectroscopy of 6-Dodecanoyl-N,N-dimethyl-2-naphthylamine (Laurdan), 1,6-diphenyl-1,3,5-hexatriene (DPH) and Di-8-ANEPPS is performed on large unilamellar vesicles (LUVs). LUVs are formed by means of the extrusion method as described in [26]. The lipid and the indicated fluorescent probe are dissolved and mixed in chloroform/methanol (1:1 v/v). Laurdan or DPH are mixed with the lipids in the initial organic solution at 1:200 probe:lipid molar ratio. The fluorescent probe di-8-ANEPPS in ethanol at 1 mg/mL stock solution concentration, is mixed with the lipids in the initial organic solution at 1:250 probe-to-lipid molar ratio. Afterward, the solvent is removed under a stream of oxygen-free dry nitrogen. The residues are subsequently maintained under vacuum overnight, and then filtered (0.2 µm) sugar solution is added at room temperature (22 °C) to yield a lipid concentration of 1 mM. The samples are heated at 60°C for 5 min, vortexed for 1 min, then left in a sonication bath for 1 min, and finally cooled in ice for

5 min. This heating/cooling procedure is repeated three times to ensure the sample homogenization. The multilamellar vesicles obtained at this stage are then extruded with a LiposoFast small-volume extruder equipped with polycarbonate filters (Avestin, Ottawa, Canada) as follows: 11 extrusions through 800 nm, followed by 21 extrusions through 100 nm filters. The final lipid concentration in cuvette is 200 µmol/L. LUV samples are studied the same day after equilibration for 20 min at room temperature (22 °C). Each sample is measured 10 times after gentle pipetting and averaged by three different LUV preparations.

2.2.3. Preparation of Bilayer Lipid Membranes

Bilayer lipid membranes (BLMs) are prepared after Montal and Mueller [27,28] in BC-20A chamber (Eastern Scientific LLC, Rockville, MD, USA), consisting of two similar Teflon blocks, each one with volume 2 mL, designed to hold 0.025 mm thin Teflon film in between. Solvent-free membrane patch is generated across a hole in the Teflon film with diameter 100 µm following the procedure described in [27]. The first step consists in pre-painting each side of the hole with ~1 µL of 1 wt% hexadecane in pentane and allowing the solvent to evaporate for ~5 min. The next step consists in pipetting in each compartment 0.5 mL of the aqueous solution containing 1 mM NaCl for the control sample or 1 mM of NaCl and 200 mM of glucose, fructose or sucrose. Thereafter, 15 µL of freshly prepared membrane-forming lipid solution is spread on top of the buffer in both compartments in order to ensure the presence of several lipid layers on the air-water interphase. The gentle addition of 1.5 mL of the corresponding aqueous solution consequently in each part of the cuvette leads to the generation of POPC bilayer on the aperture of the Teflon partition between the two compartments [28]. The final control of membrane formation and quality is performed electrically [27]. EIS-FTT measurement is effectuated immediately afterward.

2.2.4. Fluorescence Spectroscopy of Laurdan-, DPH- and Di-8-ANEPPS-Labeled LUVs

The lipid structural order parameter of the bilayer in the fatty acid core and at the glycerol level are probed by fluorescence spectroscopy of DPH [29] and Laurdan [30], respectively. The excitation wavelength for Laurdan is 355 nm. In disordered lipid surroundings, its emission maximum with intensity I_{490} is centered at 490 nm. In more ordered lipid environment, the maximum of Laurdan emission is shifted at 440 nm with intensity I_{440}. The lipid packing is quantified by the parameter GP representing Laurdan generalized polarization accordingly [30]:

$$GP = \frac{I_{440} - I_{490}}{I_{440} + I_{490}} \tag{1}$$

It theoretically assumes values from -1, corresponding to disordered membrane up to 1 for most ordered molecules. We record three times all emission spectra from 390 to 600 nm, then average, and subtract background. For every studied LUV suspension, the emission spectrum of Laurdan is obtained and GP values are calculated in the temperature range $(20 - 60)$ °C.

We apply 1,6-diphenyl-1,3,5-hexatriene (DPH) fluorescence spectroscopy [31] to assess membrane fluidity as a function of sugar concentration in LUV samples. DPH represents a hydrocarbonic probe, which is nearly non-fluorescent in water. Contrastingly, in lipid membranes, it exhibits strong fluorescence. In our experiment, the excitation is adjusted to 358 nm. We record the emission at 430 nm. The fluorescence polarization is measured as described in [32]. The DPH fluorescence anisotropy r_{DPH}, quantifying membrane fluidity is expressed by the intensities I of the polarized components of the fluorophore emission:

$$r_{DPH} = \frac{I_{VV} - GI_{VH}}{I_{VV} + 2GI_{VH}}, \tag{2}$$

where V and H stand for "vertical" and "horizontal", corresponding to the orientation of the polarization axis with respect to the light direction. The first index indicates the polarization of the excitation light and the second one denotes the polarization of the emission signal, corresponding to the orientation of the excitation and emission polarizers, respectively. The grating factor $G = I_{HV}/I_{HH}$ reflects the sensitivity of the instrument toward vertically and horizontally polarized light. Around the main phase transition of the bilayer, the fluorescence anisotropy of DPH sharply increases. It can assume values from -0.2 to 0.4 [31].

The membrane dipole potential in sugar-containing aqueous solutions is quantified by the fluorescence excitation ratio of the potential-sensitive fluorescent styryl dye di-8-ANEPPS. In the case of di-8-ANEPPS incorporated into LUVs, fluorescence is excited at 420 nm and 520 nm and detected at 670 nm. It has been shown that the fluorescence intensity ratio $R_{ex} = I_{670\,(exc.420)}/I_{670\,(exc.520)}$ is proportional to the dipole potential Ψ_d, independently of fluidity effects [33–35]:

$$\Psi_d = (R_{ex} + 0.3)/0.0043 \tag{3}$$

The dipole potential occurs transversally between the water–lipid interface and the hydrocarbon interior from the contribution of all polarized and polarizable chemical groups of lipid molecules and by the hydration shell of the bilayer [36].

FP-8300 spectrofluorimeter (Jasco) equipped with polarizers and a thermostatted ($\pm0.1\ ^{\circ}$C) cuvette holder with quartz cuvettes is used. Samples are equilibrated for 5 min at the desired temperature. Excitation and emission slits are adjusted to 5 nm.

2.2.5. Electrodeformation of GUVs

The specific electrical capacitance, C_m, of POPC and SOPC membranes in the presence of small carbohydrates (sucrose, glucose and fructose) is measured from the frequency-dependent deformation of GUVs in alternating electric field [37]. The vesicle shape transformation has been established to depend on the ratio Λ between the conductivity λ_{in} of the aqueous solution, enclosed by the vesicle membrane, and the conductivity λ_{out} of the external (suspending) medium [38,39]. In the case of more conductive external medium, upon increasing the AC field frequency, a vesicle with radius r placed in a more conductive aqueous solution changes its shape from prolate to oblate with respect to the field direction. During this morphological transition, the intermediate frequency, f_{cr}, at which the quasispherical shape is assumed, is given by the expression [38]:

$$f_{cr} = \frac{\lambda_{in}}{2\pi r \overline{C}_m}[(1 - \Lambda)(\Lambda + 3)]^{-1/2} \tag{4}$$

$\overline{C}_m$ denotes the resultant capacitance of a series of three capacitors. These are the bare lipid bilayer, C_m, and the capacitances of the diffuse charge regions resembling electric double layers in the aqueous solution at the two sides of the bilayer, denoted by $C_{D,in}$ and $C_{D,ex}$, respectively:

$$\overline{C}_m = (1/C_m + 1/C_{D,in} + 1/C_{D,ex})^{-1} \tag{5}$$

On the length scale of GUVs with radii $\sim$10 μm and bilayer thickness $d \sim 5$ nm ($d \ll r$) [37] the membrane is described as a two-dimensional surface with dielectric permittivity $\varepsilon_m = \varepsilon_{rm}\varepsilon_0$, where ε_{rm} stands for the relative dielectric constant of the bilayer and $\varepsilon_0 \approx 8.85 \times 10^{-12}$ F/m is the vacuum permittivity. Therefore, the specific capacitance C_m of the bilayer is given by:

$$C_m = \varepsilon_m/d \tag{6}$$

The capacitance of the electric double layers is represented by the capacitance of a planar capacitor with thickness equal to the Debye length, λ_D, and dielectric constant equal to the dielectric constant of the aqueous solution $\varepsilon_r \approx 80$ [40]. The Debye length is related to the molar concentration c of a 1:1 electrolyte by the expression $\lambda_D = 0.303/\sqrt{c}$ nm [41]. Considering the concentrations of NaCl applied here, we estimate the capacitance for the double layers of free charges in the aqueous solution on both sides of the bilayer $C_{D,in}$ and $C_{D,ex}$. As discussed in [38], their contribution increases at lower salt concentrations as well as for high enough values of the bilayer capacitance.

It has been established that the dielectric properties of the ionic double layers near the membrane are related to the orientational ordering of water dipoles in the aqueous surroundings as well as in the headgroup region of lipid bilayers. We analyze the former in the light of the theoretical evaluations published so far [40,42,43], while the latter corroborates the necessity to investigate the effect of sugar molecules on the bilayer dielectric properties. In the case of lipid membranes composed of zwitterionic phosphatidylcholines, studied here, it has been shown that at much higher monovalent salt concentrations the relative permittivity in the dipolar headgroup region is decreased as a result from the saturation effect in orientational ordering of water dipoles [44]. In the present study, we consider PC membranes in sugar-containing electrolyte solutions with ionic strength, which is two orders of magnitude lower than in [44]. At zero surface charge density corresponding to the model lipid system studied here, the value taken for $\varepsilon_r \approx 80$ represents a good evaluation for the relative permittivity of the aqueous solution surrounding the bilayer as shown in [43]. The electric field strengths applied are far below the electroporation threshold [37,45].

The electrodeformation measurements are conducted in a chamber consisting of two parallel glass slides, which are separated by a 0.5 mm-thick inert spacer (Sigma-Aldrich Inc., St Louis, MO, USA). AC electric field from an arbitrary waveform generator (33120A, HP/Agilent, Santa Clara, CA, USA) is applied to a pair of two rectangular parallel ITO-electrodes deposited on the lower inner surface of the chamber at 1 mm apart. The measurement is performed by varying the frequency of the imposed uniform field in the range of 10–200 kHz. The field strengths ≤ 7 kV/m, applied here, are two orders of magnitude lower than the electroporation threshold [37,45]. The vesicle electrodeformation is observed and recorded using a phase-contrast microscope (B-510PH, Optika, Ponteranica, BG, Italy) equipped with a dry objective ($\times 40$, 0.65 numerical aperture) and Axiocam ERc 5s camera 5 MP (Zeiss, Jena, Germany) connected to a computer for image recording and processing with resolution of 0.1 μm/pixel. The ratios $0.87 \leq \Lambda \leq 0.95$ correspond to a more conductive suspending solution. We perform data analysis following the original approach of Salipante et al. [16,38].

2.2.6. Fast Fourier Transform Impedance Spectroscopy of BLMs

Fast Fourier transform electrochemical impedance spectroscopy (FFT-EIS) is based on measurements in the time domain, applying a multisine perturbation signal of a small-amplitude ~10 mV peak-to-peak covering the desired frequency range of 1.5 Hz–50 kHz. The perturbation and the respective response signals are simultaneously measured and subsequently transferred to the frequency domain using the fast Fourier transformation (FFT). The major advantages of the approach chosen here are the fast acquisition of the impedance spectrum obtained in a couple of seconds and the possibility to easily recognize violation of stationarity [46,47]. Thus, the FFT-EIS method appears to be suitable for studying the impedance properties of dynamic and unstable samples such as lipid bilayers. The perturbation voltage is applied potentiostatically using a two-electrode scheme with two platinum electrodes to supply the current and to measure the voltage across membrane. The contact area of Pt electrodes with the bulk phase in both compartments of Montal–Mueller cell is large enough to guarantee that the contribution of the electrode double electric layer is negligible in the frequency range covered.

The analysis of FFT-EIS data is performed by equivalent circuit modeling. From electrical point of view Montal–Mueller BLM set-up is represented by the capacitance C_{BLM} and resistance R_{BLM} of the planar lipid bilayer, connected in parallel together with the capacitance of the Teflon membrane C_{TM}. The resistance R_s of the surrounding solution is connected in series as depicted in Figure 1. Considering the 100 μm aperture surface area $S_m = 8.10^{-5}$ cm^2 we deduce membrane-specific capacitance $C_m = C_{BLM}/S_m$ and resistance $R_m = R_{BLM}S_m$. The reported values are calculated as the weighted average of 4–7 independent measurements averaged over 10 repetitions each. The capacitance C_{TM} is measured independently as discussed below.

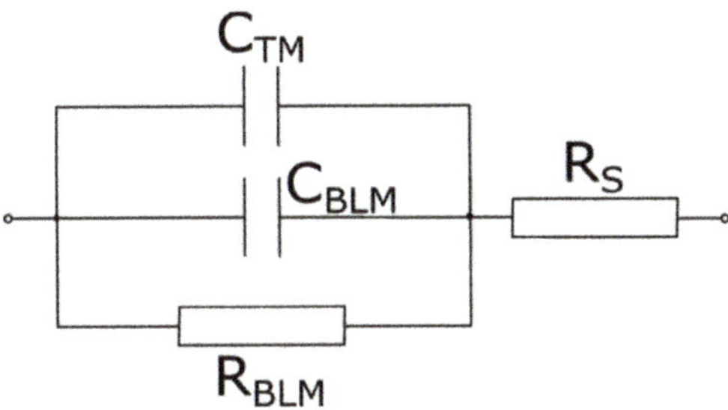

Figure 1. Equivalent circuit model of Montal–Mueller cell for formation of bilayer lipid membranes (BLMs): C_{BLM}—BLM capacitance; R_{BLM} —BLM resistance; C_{TM} —capacitance of the Teflon membrane; R_s —resistance of the aqueous solution.

3. Results

3.1. Specific Capacitance of Lipid Bilayers in the Presence of Simple Carbohydrates

3.1.1. Free-Standing Lipid Bilayers—Electrodeformation of GUVs

The specific capacitance of POPC bilayers is measured in aqueous solutions of 1 mM NaCl containing different concentrations of glucose, fructose or glucose up to 300 mmol/L. In order to probe for specific effects, we determine the same parameter also for another type of phosphatidylcholine membranes, namely SOPC, at 200 mmol/L of sucrose in the aqueous surroundings. Our choice of synthetic lipids is motivated by their controlled chemical composition combined with the property to mimic well natural lipid extracts, e.g., egg-yolk or soybean phosphatidylcholine extracts. The values obtained for membrane specific capacitance together with the error calculated from the fit of our experimental data for every sugar concentration are summarized in Table 1 and Figure 2.

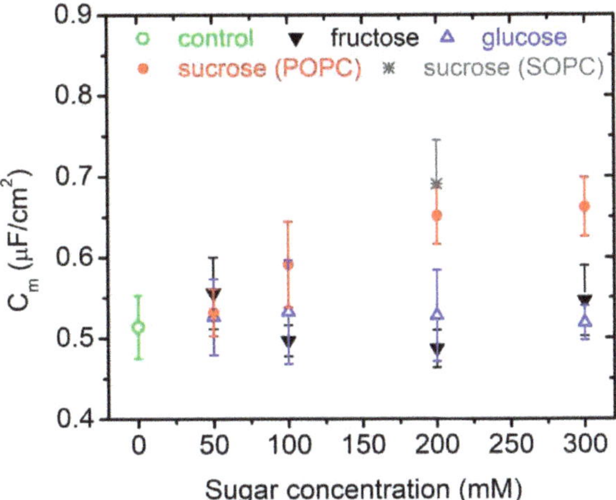

Figure 2. Specific capacitance of (i) POPC bilayers in the presence of 1 mmol/L NaCl and up to 300 mmol/L of glucose, fructose or sucrose; (ii) SOPC membranes in 1 mmol/L NaCl, 200 mmol/L of sucrose obtained from electrodeformation of GUVs.

For each sugar concentration, we measure the transition frequency f_{cr} (Equation (4)) of 6 to 15 vesicles. The acquired experimental data for f_{cr} as a function of the inverse radii of vesicles are fitted for fixed λ_{in} and Λ. Latter are the same for all recorded and analyzed vesicles in one batch. Membrane capacitance serves as a single fitting parameter. Our results show no alteration of the specific capacitance of POPC bilayers in aqueous solutions of glucose and fructose at concentrations of up to 300 mmol/L and in the presence of 1 mmol/L NaCl. In glucose solutions, we obtain $C_m = 0.52 \pm 0.02$ µF/cm^2 with goodness of fit 0.91. The value calculated from the data acquired in fructose solutions is $C_m = 0.51 \pm 0.01$ µF/cm^2 with goodness of fit 0.55. In sucrose-containing aqueous surroundings we measure 25%-higher values of membrane specific capacitance at 200 and 300 mmol/L sugar concentrations compared to the control value of the same parameter in 1 mmol/L NaCl (Figure 2). Similar increase of C_m is reported also for SOPC bilayers at 200 mmol/L of sucrose in water.

Table 1. Specific capacitance of POPC membranes in aqueous solutions containing different concentrations up to 300 mmol/L glucose, fructose or sucrose; data obtained from electrodeformation of GUVs; GF, goodness of fit. The errors in $\overline{C}_m$ and C_m are standard deviations.

Sugar, mmol/L	λ_{in}, µS/cm	Λ	$\overline{C}_m$, µF/cm^2 (Number of Vesicles)	C_m, µF/cm^2	GF
			Control		
0	258	0.87	0.44 ± 0.03 (7)	0.51 ± 0.04	0.75
			Sucrose		
50	276	0.95	0.45 ± 0.02 (9)	0.53 ± 0.03	0.73
100	238	0.93	0.49 ± 0.04 (11)	0.59 ± 0.05	0.44
200	363	0.95	0.53 ± 0.03 (10)	0.65 ± 0.04	0.82
300	347	0.94	0.54 ± 0.03 (8)	0.66 ± 0.04	0.84
			Glucose		
50	318	0.94	0.45 ± 0.04 (8)	0.53 ± 0.05	0.93
100	257	0.88	0.45 ± 0.05 (6)	0.53 ± 0.06	0.66
200	192	0.88	0.45 ± 0.05 (6)	0.53 ± 0.07	0.31
300	223	0.90	0.44 ± 0.02 (8)	0.52 ± 0.02	0.96
			Fructose		
50	145	0.88	0.47 ± 0.04 (15)	0.56 ± 0.04	0.53
100	312	0.95	0.43 ± 0.02 (7)	0.50 ± 0.02	0.48
200	148	0.87	0.42 ± 0.02 (10)	0.49 ± 0.02	0.80
300	145	0.87	0.45 ± 0.04 (10)	0.55 ± 0.04	0.64

3.1.2. Suspended Planar Lipid Bilayers—FFT-EIS of BLMs

Fast Fourier transform electrochemical impedance spectroscopy (FFT-EIS) is applied to probe the effect of simple carbohydrates on the impedance characteristics of POPC bilayer lipid membranes. Montal–Mueller technique for planar lipid bilayers yields stable bilayer membranes for assessment of their impedance properties to be probed here by an FFT electrochemical impedance spectrometer. Control samples are obtained in 1 mmol/L NaCl. Impedance measurements are performed also in the presence of 1 mmol/L NaCl and 200 mmol/L glucose, fructose or sucrose. Solvent-free BLMs are produced by Montal–Mueller technique [27,28] across a 100 µm aperture in a 0.025 mm thin Teflon film. The application of Montal–Mueller cell for the measurement of BLM electrical properties requires considering the capacitance C_{TM} of the Teflon membrane. The value measured for $C_{TM} = 56.33 \pm 0.23$ pF is comparable to the expected capacitance of the bilayer lipid membrane suspended on the hole. Hence, it has been considered in the equivalent circuit modeling as shown in Figure 1. In order to exclude any preparation-related artifacts, the impedance of at least four POPC BLMs is measured for each carbohydrate tested. Impedance spectra plots in the complex plane (Nyquist diagrams) measured for POPC BLMs in 1 mmol/L NaCl (solid circles) and 200 mmol/L of sucrose (hollow circles) with equivalent model circuit fits (lines) are displayed in Figure 3. The average values of the specific capacitance and resistance of POPC BLMs with their standard deviations obtained in 1 mmol/L NaCl aqueous solution and in 200 mmol/L sugar-containing media are summarized in Table 2 and in Figure 4.

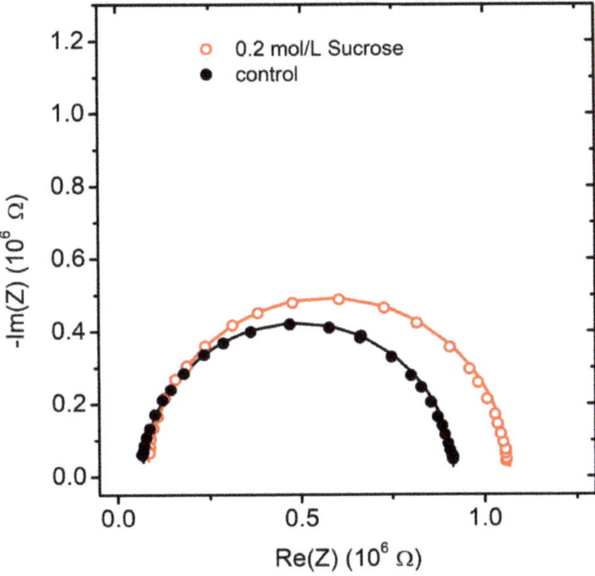

Figure 3. Impedance spectra plots in the complex plane (Nyquist diagrams) measured for POPC BLMs in 1 mmol/L NaCl (solid circles) and 200 mmol/L of sucrose (hollow circles). Lines represent equivalent model circuit fits.

Table 2. Specific capacitance and resistance of POPC membranes in aqueous solutions containing 1 mmol/L NaCl and 200 mmol/L glucose, fructose or sucrose; data obtained from FFT-EIS of BLMs by equivalent model circuit fits; control samples contain only 1 mmol/L NaCl.

Sugar	R_m, 10^6 Ω cm^2	C_m, μF/cm^2	Number of Samples	GF
Control	1.42 ± 0.05	0.79 ± 0.06	7	0.51
Glucose	1.62 ± 0.08	0.85 ± 0.07	4	0.16
Fructose	1.57 ± 0.06	0.87 ± 0.05	7	0.71
Sucrose	1.56 ± 0.02	1.10 ± 0.10	6	0.43

All measurements are performed in aqueous solutions containing 1 mmol/L NaCl. The equivalent model circuit fits give for the studied samples values of the aqueous surrounding series resistance R_s (Figure 1) ranging from 36 ± 1 to 136 ± 8 kΩ. In all aqueous solutions, we measure POPC membrane resistance $R_m \approx 10^6$ Ω cm^2 (cf. Table 2), which is similar to the values for lipid bilayers composed of other synthetic phosphatidylcholines as reported so far [48–50]. Upon addition of sugars in the aqueous phase we observe a slight increase of ~10–14% in BLM resistance as shown in Table 2.

FFT-EIS measurements of POPC BLMs impedance yield higher capacitance values compared to C_m deduced from experiments with GUVs for the same lipid bilayer and aqueous solution compositions (cf. Tables 1 and 2). Membrane specific capacitance remains unchanged in 200 mmol/L glucose or fructose solutions, while in the presence of the same concentration of sucrose, we obtain capacitance value almost 40% higher compared to the sugar-free control sample (cf. Table 2 and Figure 4). The latter is qualitatively consistent with the respective results for free-standing POPC membranes acquired from electrodeformation of GUVs, which have shown slightly lower increase of 25% of membrane specific capacitance.

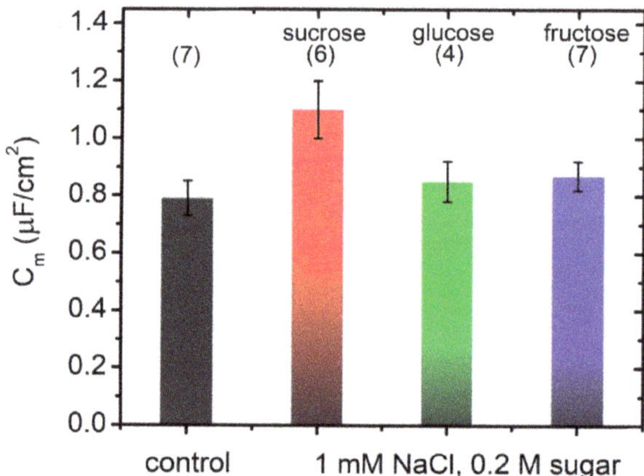

Figure 4. Specific capacitance of POPC BLMs in 1 mmol/L NaCl (control); 1 mmol/L NaCl and 200 mmol/L of sugar (sucrose, glucose and fructose); data acquired by FFT-EIS; C_m values calculated as the weighted average of 4–7 independent measurements, each of them averaged over 10 repetitions.

3.2. Lipid Packing in the Presence of Sugars

We probe the effect of glucose and sucrose at concentrations in the aqueous phase up to 400 mmol/L on the lipid packing in the bilayer at various temperatures ranging from 20 to 60 °C. Normalized fluorescence emission spectra of Laurdan/SOPC and Laurdan/POPC vesicles at different temperatures in bidistilled water as well as in 400 mmol/L glucose and sucrose aqueous solutions are depicted in Figure 5. The emission spectra at 20 °C of both control and sugar-containing POPC and SOPC vesicles exhibit two peaks of nearly equal intensities centered at 430 nm (blue-shifted) and 490 nm (red-shifted), respectively. The presence of two peaks indicates that Laurdan senses two environments, one ordered and another one, more disordered, which could be associated with the presence of saturated and unsaturated fatty acids in PC molecules.

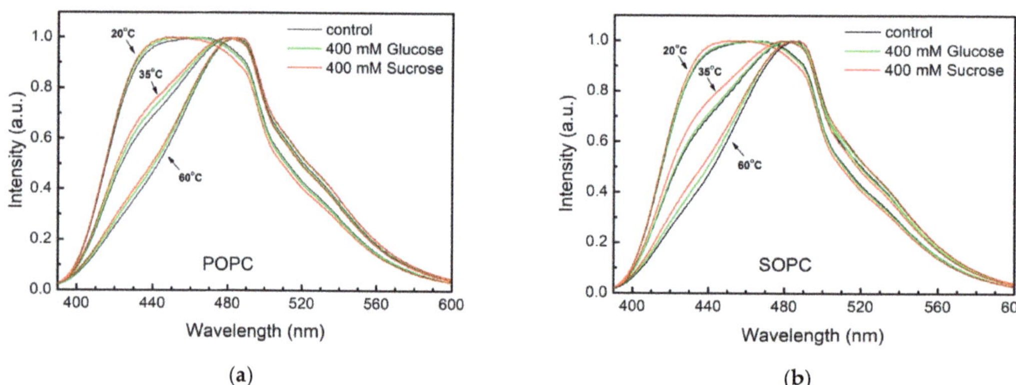

Figure 5. Laurdan intensity profiles from (**a**) POPC and (**b**) SOPC LUVs in water (control) and in 400 mmol/L glucose or sucrose scanned at 20, 35, and 60 °C.

At 20 °C the spectral profiles of POPC vesicles in sugar solutions resemble the profile of the control sample similarly characterized by two peaks centered at 430 and at 490 nm. At the same temperature, the disordered red-shifted peak is characterized with slightly lower intensity compared to the blue-shifted one in sucrose solutions both for POPC and SOPC vesicles (20 °C, red curves in Figure 6a,b). The temperature increase leads to the progressive disappearance of the more ordered blue-shifted peak reflecting that at 60 °C the mobility of the two fatty acids is undistinguishable for the fluorescent probe.

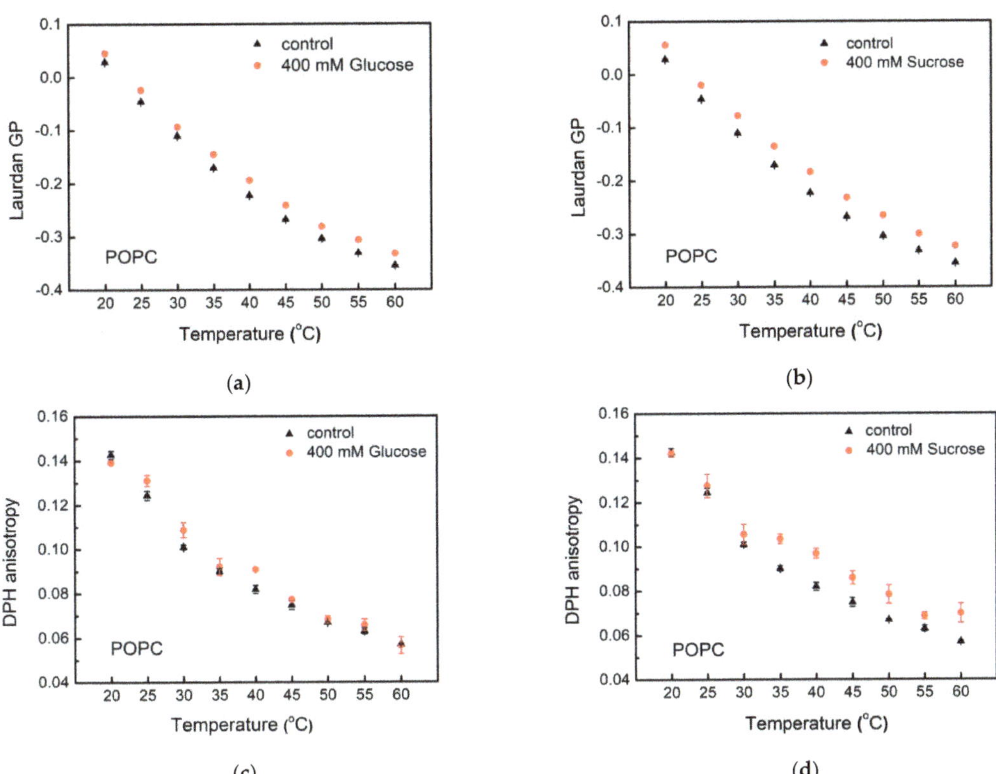

Figure 6. Fluorescence spectroscopy data for POPC membranes: Laurdan *GP* as a function of temperature in: (**a**) glucose and (**b**) sucrose solutions; DPH anisotropy vs. temperature in (**c**) glucose and (**d**) sucrose solutions.

Laurdan *GP* values for POPC LUVs in water (control) and in sugar-containing aqueous solution as a function of the temperature are presented in Figure 6a,b. For the sake of clarity only the data for the highest sugar concentration studied are displayed in the figures. The fluorescence emission spectrum of Laurdan-labeled LUVs is measured at 20 to 60 °C. The generalized polarization *GP* is calculated according to Equation (1) in the control sample as well as for the carbohydrate concentrations studied. As it is indicated above, *GP* scale values vary between −1 and 1. In our experiments, the control POPC LUVs exhibit *GP* values from 0.05 (20 °C) to −0.35 (60 °C) corresponding to intrinsically loosely packed lipids in liquid-disordered phase (L_d). A similar trend is observed for Laurdan *GP* values for POPC LUV suspensions in the presence of glucose and sucrose. In sugar-containing aqueous environment, we report higher *GP* values compared to the control samples in the whole temperature range scanned.

We quantify the effect of glucose and sucrose on the molecular organization in membrane hydrophobic core using DPH. The fluorescence anisotropy of this molecular probe is related to the fatty acids mobility [31]. The steady-state DPH anisotropy, r_{DPH}, within the bilayer is determined according to Equation (2). As discussed above, the scale values of the parameter vary between −0.2 and 0.4. Here, DPH fluorescence anisotropy changes from 0.14 to 0.06 for POPC and SOPC vesicles in water (Figures 6 and 7), thus indicating membranes in liquid-disordered phase in the temperature range studied from 20 to 60 °C.

This representation allows for discerning the different trends in the thermotropic behavior of POPC and SOPC bilayers in glucose and sucrose solutions.

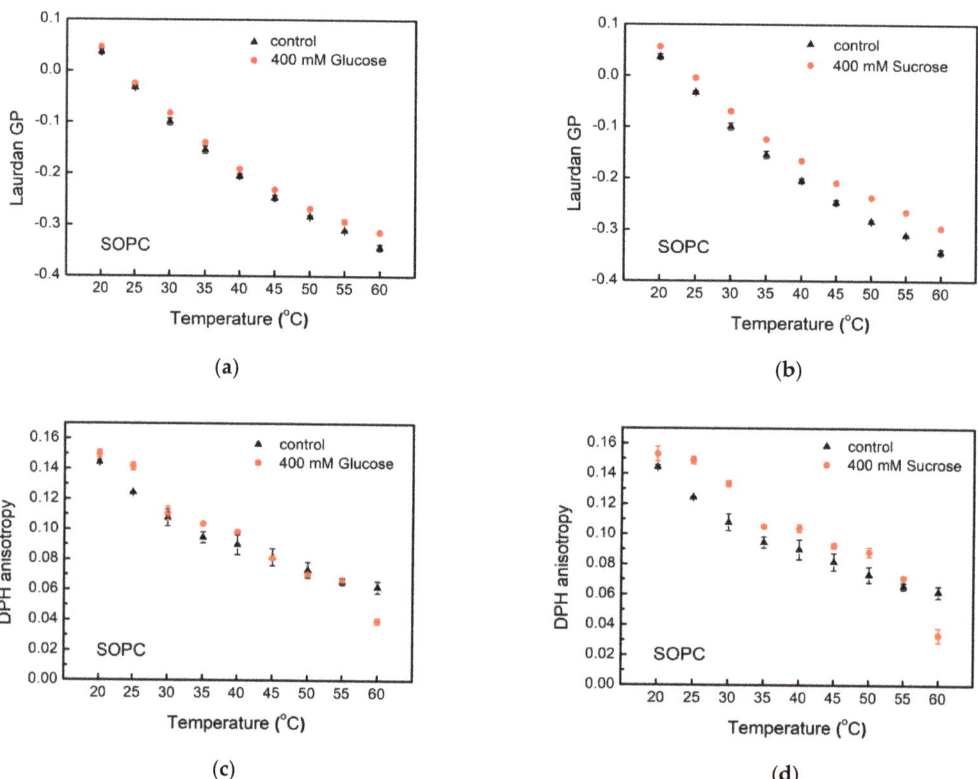

Figure 7. Fluorescence spectroscopy data for SOPC bilayers: Laurdan *GP* as a function of temperature in: (**a**) glucose and (**b**) sucrose solutions; DPH anisotropy vs. temperature in (**c**) glucose and (**d**) sucrose solutions.

Upon increasing temperature, DPH fluorescence anisotropy is reduced both in POPC and SOPC LUVs, which corresponds to an increase in membrane fluidity. For the two PC studied, we observe different behavior of r_{DPH} upon addition of glucose and sucrose in the aqueous phase. While in glucose solutions, no correlation between DPH fluorescence anisotropy of POPC vesicles and the monosaccharide concentration is found, the presence of sucrose in the aqueous surroundings leads to a decrease in DPH rotational diffusion. As far as DPH anisotropy is inversely proportional to membrane fluidity, the above results correspond to the formation of more ordered liquid hydrocarbon region of POPC bilayers in the presence of sucrose compared to the control sample. In sucrose solutions with increasing the temperature the value of DPH fluorescence anisotropy for both lipid compositions decreases differently in comparison to r_{DPH} reduction of the control sample in water upon heating (Figures 6d and 7d). At 20 °C DPH anisotropy in POPC membranes is lower in glucose-containing solution than in water, which corresponds to higher membrane fluidity. Inverse effect of sucrose on r_{DPH} is observed at higher temperatures corresponding to decreased rotational diffusion of the fluorophore (Figure 6c,d). For SOPC, the inverse picture is observed with higher fluorescence anisotropy at 20 up to 55 °C. Upon further heating, we measure increased fluidity of SOPC bilayers both in glucose and sucrose solutions compared to the control sample (Figure 7c,d).

A noteworthy finding is the qualitatively different behavior of DPH anisotropy in POPC and SOPC bilayers as shown in Figures 8 and 9. In Figure 8, fluorescence spectroscopy data for single-component POPC and SOPC vesicles in water and in 400 mmol/L sugar solutions (glucose and sucrose) are shown as the difference Δ Laurdan GP between GP at 60 and 20 °C, and the difference Δ DPH anisotropy between DPH fluorescence anisotropy at 60 and 20 °C, respectively. The change of lipid ordering and membrane fluidity of PC vesicles in water and sugar-containing solutions are displayed as the temperature increases. Sucrose solutions membranes (especially SOPC ones) are more thermostable compared to controls because of the smaller change between GP at 60 and 20 °C. Upon increasing the temperature, the effect of sucrose on the hydrophobic core fluidity is more pronounced for SOPC vesicles, while for POPC membranes, the fluidity alteration is weaker.

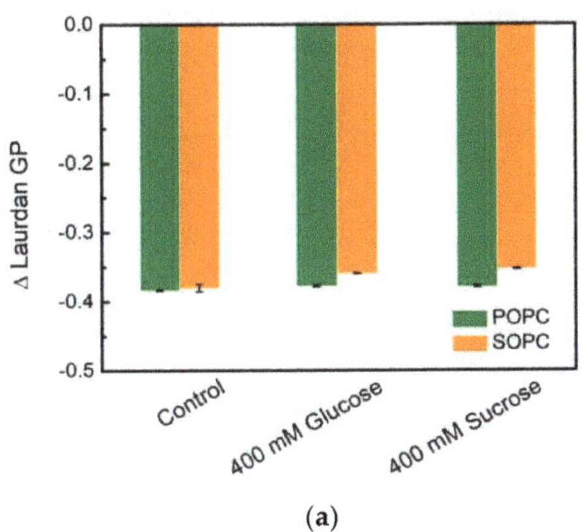

(a)

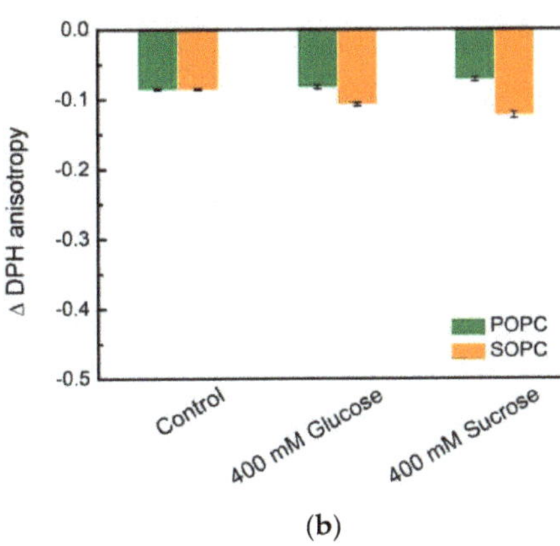

(b)

Figure 8. Lipid packing and membrane fluidity in single-component POPC and SOPC vesicles in water (controls) and in 400 mmol/L sugar solutions (glucose and sucrose): (**a**) Δ Laurdan GP is defined as the difference between GP at 60 and 20 °C; (**b**) Δ DPH anisotropy is defined as the difference between anisotropy at 60 and 20 °C. DPH anisotropy and Laurdan GP values represent the mean of three independent experiments. Error bars represent standard errors.

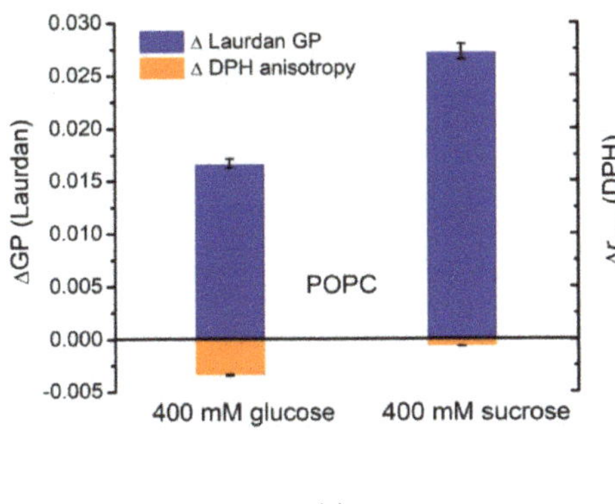

(a)

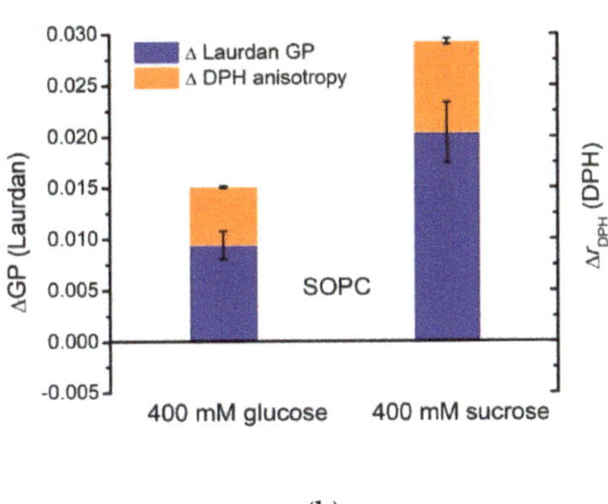

(b)

Figure 9. ΔGP and Δr_{DPH} calculated as the difference between DPH anisotropy or Laurdan GP of (**a**) POPC and (**b**) SOPC vesicles in 400 mmol/L glucose or sucrose and in water (control) at 20 °C. DPH anisotropy and Laurdan GP values represent the mean of three independent experiments. The error bars correspond to the standard deviations.

Figure 9 represents ΔGP and Δr_{DPH}, calculated as the difference between Laurdan generalized polarization and DPH anisotropy, respectively, measured for POPC vesicles with and without sugars in the bulk phase at 20°C. The same quantities are calculated also for SOPC LUVs (Figure 9b). The plot allows for discerning the different strength of the effect for the two types of PC studied in the presence of the mono- and disaccharide. The comparison between the reported values of Δr_{DPH} and ΔGP suggests that the presence of glucose or sucrose in the aqueous surroundings affects the lipid packing in the bilayer more strongly at the glycerol level for both lipid compositions displayed in Figure 9. For SOPC bilayers in 400 mmol/L sucrose solutions, we obtain $\Delta GP/\Delta r_{DPH} \sim 2.3$ compared to ~ 1.7 at the same concentration of glucose. The effect is more pronounced for POPC samples yielding $\Delta GP/\Delta r_{DPH} \sim -43$ in sucrose-containing environment and ~ -5 in glucose solutions, respectively. Here, the negative values reflect the slight reduction of DPH anisotropy in POPC membranes upon addition of 400 mmol/L glucose or sucrose in the aqueous surroundings (Figures 8 and 9).

3.3. Dipole Potential in Lipid Bilayers and the Effect of Simple Carbohydrates

The fluorescence intensity ratio R_{ex} of di-8-ANEPPS dye (Equation (3)) in single-component lipid membranes composed of POPC and SOPC is measured in order to explore possible alterations in the primary hydration shell of the bilayer in the presence of sugars [51]. As discussed above, the measured value of R_{ex} is related to the electrical potential, which is generated transversally at the water-lipid interface and the hydrocarbon interior as a result of the contribution of polarized and polarizable chemical groups of lipid molecules and water surrounding the bilayer [36].

For both types of phospholipids, we obtain an increase in di-8-ANEPPS fluorescence intensity ratio in 1 mmol/L NaCl and 200 mmol/L of the studied sugars compared to its value measured in water. The increase of R_{ex} stands for an increase of membrane dipole potential following from the interaction of solutes in the aqueous phase and the lipid bilayer. Slight alterations in membrane fluorescence intensity ratio are measured for POPC and SOPC bilayers upon the addition of 1 mmol/L NaCl and 200 mmol/L of glucose, fructose or sucrose (Table 3 and Figure 10).

Table 3. Di-8-ANEPPS fluorescence intensity ratio and dipole potential of POPC bilayers in aqueous solutions containing 1 mmol/L NaCl and 200 mmol/L glucose, fructose or sucrose; control samples are measured in bidistilled water and in 1 mmol/L NaCl; fluorescence excited at 420 and 520 nm and detected at 670 nm.

Sample	POPC		SOPC	
	R_{ex}	Ψ_d (mV)	R_{ex}	Ψ_d (mV)
H_2O, bidistilled	1.798 ± 0.002	488	1.789 ± 0.002	486
1 mM NaCl	1.800 ± 0.002	488	1.794 ± 0.001	487
200 mM Glucose, 1 mM NaCl	1.806 ± 0.004	490	1.797 ± 0.002	488
200 mM Fructose, 1 mM NaCl	1.804 ± 0.002	489	1.795 ± 0.003	487
200 mM Sucrose, 1 mM NaCl	1.807 ± 0.002	490	1.796 ± 0.001	487

The dipole potential measured here for POPC and SOPC membranes with and without sugars is consistent with the values obtained by other authors and methods in the range of 200–1000 mV for biomembranes [52,53]. The dipole potential of PC membranes is reported around 410 ± 150 mV in dependence of their degree of saturation, polar heads and the physicochemical properties of the buffer used. At low pH, as well as in the presence of molecules that increase the molecular order parameter and decrease the rotational order parameter, lipid membranes are susceptible to further increase of the dipole potential, similar to the effect of saturated PC compared to unsaturated PC.

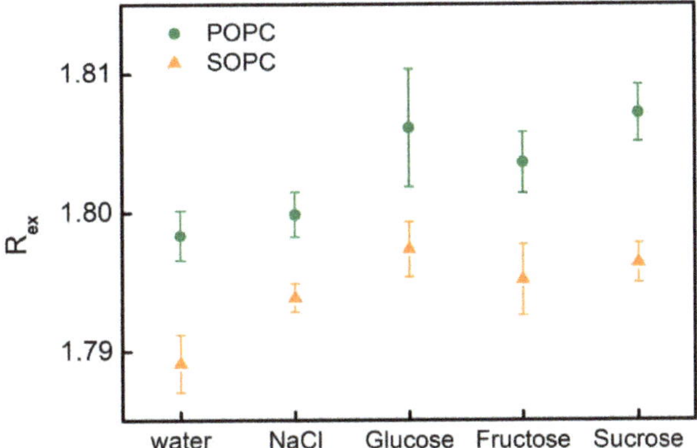

Figure 10. Di-8-ANEPPS fluorescence intensity ratio of POPC (circles) and SOPC (triangles) bilayers in 1 mmol/L NaCl sugar-free solutions and 200 mmol/L of glucose, fructose or sucrose; fluorescence excited at 420 and 520 nm and detected at 670 nm. Values are the mean of three independent experiments; the error bars correspond to the standard deviations.

4. Discussion

Both experimental methods applied in the present study for the measurement of the membrane electrical capacitance yield higher C_m values at 200 mmol/L of sucrose in the aqueous surroundings compared to its value in the sugar-free control sample. The capacitance obtained from measurements on GUVs in sugar-free aqueous solutions $C_m = 0.51 \pm 0.04 \, \mu F/cm^2$ is lower than values $\sim 0.55 \div 1 \, \mu F/cm^2$ measured for planar bilayers of different lipid composition and charge [28,50,54–57] (data overview in Ref. [58]). In the present study, we confirm this trend by reporting higher specific electrical capacitance from experiments on Montal–Mueller BLMs compared to the values acquired from GUVs. Fluctuating vesicles are characterized by low-tension membranes [59,60], whereas for planar lipid bilayers the lateral tension is orders of magnitude higher [61,62]. Thus, a reduced membrane thickness of BLMs compared to GUVs is expected due to the tension difference for the two bilayer systems. Patch-clamp experiments on GUVs recently explored the bilayer thinning effect on membrane capacitance in 200 mmol/L sucrose and glucose solutions [58]. They have shown that membrane capacitance can vary with tension by up to 3%. The measurements reported in [58] have been performed in the relatively high-tension regime of mN/m. In the present study, we employed fluctuating free-standing vesicles with low membrane tensions, $10^{-6} \div 10^{-4}$ mN/m as deduced from fluctuation spectroscopy [59,63–65]. Considering the vesicle electrodeformation involved in the capacitance measurements, here, membrane tensions are slightly elevated to 10^{-3} mN/m [39,66,67], which is orders of magnitude lower than the mN/m tensions applied in [58]. Despite the different tension range in [58], it can be concluded that membrane thinning associated to tension only partially explains the differences in the capacitance data acquired from GUVs and from planar lipid membranes. In our previous study [16] providing the first experimental evidence about the influence of sucrose on the electrical properties of lipid bilayers, we explored the possibility of changes in membrane dielectric constant as a factor affecting the capacitance. Here, we relate the reported increase of the electric capacitance of the bilayer and the corresponding increase in its dielectric permittivity [16] in the presence of ≥200 mmol/L sucrose to possible alterations in membrane structure and organization induced by membrane-sugar interactions [64].

The phosphatidylcholines POPC and SOPC studied here are synthetic mixed-acyl glycerophospholipids. The monounsaturated oleic acid residue (18:1) is positioned identically in both lipids (sn-2 position), while the saturated hydrocarbon chain at sn-1 position is different, in palmitic and stearic acid for POPC and SOPC, respectively. POPC bilayers undergo a gel-to-liquid crystalline phase transition around $-2\ °C$ [65]. The main phase transition of SOPC membranes occurs around $7\ °C$ [66].

The fluorescence spectroscopy of Laurdan and DPH provide information about the lipid packing at different levels in the lipid bilayer, near the glycerol backbone and the hydrophobic core, respectively. Laurdan responds to the degree of hydration at the glycerol level, while DPH fluorescence anisotropy corresponds to the rotational diffusion of the probe in the hydrophobic region of the bilayer. Considering that the lipid molecules studied here are not characterized by truncated chains, we apply interchangeably the terms ordering and packing [67,68].

The amphiphilic fluorophore Laurdan comprises a naphthalene residue linked by an ester bond (hydrophilic) and a chain of lauric fatty acid (hydrophobic). As a result, it inserts the membrane parallel to lipid molecules in a way that its naphthalene moiety is located at the glycerol backbone of the lipid molecule and more precisely, at the level of *sn*-1 carbonyl [69]. Upon excitation by UV-light, the dipole moment of the fluorescent moiety increases leading to reorientation of the surrounding solvent dipoles. The results presented above (Figures 6 and 7) suggest that at high temperatures, membrane becomes more loosely packed, which imparts higher mobility and increased dipolar relaxation at membrane interface. The reported results indicate that at higher temperatures the Laurdan GP values decrease for all measured LUV suspension compositions. Our findings support the expected thermotropic behavior of the lipid bilayer, whose packing order decreases upon increasing the temperature.

Reorientations of *sn*-1 and *sn*-2 chains lead to conformational and hydration changes at the glycerol level related also to reorientation of the lipid headgroups. Laurdan position distribution is characterized by a certain width as well as its possible relocation upon excitation. Thereby, the fluorescence wavelength is related to the fluorophore location depth in membrane. Shorter wavelengths are emitted by Laurdan molecules positioned deeper within the lipid bilayer, while red-shifted emission (longer wavelengths) occur from fluorophores located closer to membrane-water interface. Hence, Laurdan experiments are able to distinguish POPC from SOPC lipid ordering. Our results support the hypothesis about sucrose ordering effect in membrane in the proximity of the glycerol backbone (Figure 5; blue-shifted Laurdan emission signal).

The depolarization of fluorescence has been recognized as a reliable parameter for characterization of dynamic features and thermotropic behavior of the hydrophobic regions of lipid membranes and lipoproteins [31]. Measurements of DPH fluorescence anisotropy allow evaluating the hindrance of the fluorophore mobility in the hydrophobic core of the bilayer related to alterations in packing of the aliphatic chains. We obtain that the increase of the sucrose content in the aqueous surrounding leads to hindering the DPH rotational diffusion as a result from to the formation of more ordered liquid phase.

The relative changes, Δr_{DPH} and ΔGP, state that the reduction in rotational diffusion and degree of hydration for the corresponding molecular probe is larger in the presence of the disaccharide. This reduction is more considerable for Laurdan (Figure 9). Therefore, the changes in lipid ordering induced by the presence of sucrose are predominantly at the glycerol level rather than in the hydrophobic core.

In order to further elucidate the effect of sucrose binding [64,70] on the electric properties of lipid bilayers, we study the bilayer dipole potential, which occurs due to the hydrated polar headgroups, the glycerol-ester region of the lipids and the functional group dipoles of the terminal methyl groups of hydrocarbon chains. This membrane parameter is still limitedly understood but undoubtedly recognized as an important regulator of membrane structure and function [71]. Numerous examples can be given in this respect such as the modulation of the hydration force interplaying in membrane–membrane and membrane-ligand interactions or the lipid-mediated cellular signaling in cells. The dipole potential arises in a medium over which the dielectric constant is changing in a large interval—from 2 to 80 [71]. Even if the microscopic nature of the interactions leading to the creation of the dipole potential remains not completely described, some elaborated theoretical models of its origin account for the important role of interfacial water molecules.

The slight increase in the dipole potential reported here for POPC and SOPC bilayers upon the addition of sodium chloride and small carbohydrates is in agreement with the trend measured for monolayers of dimyristoylphosphatidylcholine on the air/water interface at 20 °C, which is below the main transition temperature of the lipid [51]. Four sucrose molecules have been reported to displace three water molecules per lipid, thus producing only a weak effect on the dipole potential or the carbonyl groups in monolayers [51]. In contrast to trehalose, sucrose has not been found to interact directly with the phospholipid groups, thus implying that this disaccharide is not expected to replace water molecules in the tightly bound hydration sphere. Upon increasing the sugar concentration, the water activity in the bulk solution decreases [72]. Furthermore, the refractive index changes observed in vesicles under osmotic stress inferred alterations in the extent of hydration and/or lipid packing of phospholipid molecules [73]. Hence, sucrose would exhibit a colligative effect on fully hydrated bilayers by osmotically extruding water from the bilayer.

Our findings suggest possible alterations of the bilayer relative permittivity in sucrose-containing aqueous surroundings. They are coherent with our previous results for sucrose solutions with high ionic strength [16]. From the capacitance data and the reported membrane thickness [64], the values of the relative dielectric permittivity have been evaluated to vary from ~2.3 (sugar-free, 10 mmol/L NaCl) to ~3.5 (for sucrose concentrations above ~200 mmol/L and 10 mmol/L NaCl). Following the approach here, we estimate a narrower range of ε_{rm} changes upon increasing sucrose content in the bulk phase, characterized by an order of magnitude of lower ionic strength. POPC bilayer relative permittivity varies from ε_{rm} ~2.4 in 1 mM NaCl up to ε_{rm} ~2.9 in the presence of sucrose with concentrations >200 mmol/L in the electrolyte solution.

The above evaluations for the relative dielectric permittivity are performed considering the integral thickness of the bilayer, including its hydrophobic part and headgroup regions. In order to appropriately account for the contribution of the hydrocarbon core and the headgroup regions, the bilayer has to be described by an equivalent circuit of capacitors in series replacing C_m in Equation (5). Attempts to analyze the obtained alteration of ε_m are based on the data available in the literature for the thickness of the headgroup region of PC ~ 9 Å [74] and the thickness of dioleoylphosphatidylcholine (DOPC) bilayers in sugar-containing aqueous solutions concentrations [64]. This approximation is realistic as the head groups of DOPC and POPC are identical and both lipids are characterized by similar hydrophobic lengths. The hydrophobic thickness of DOPC and POPC is reported to be between 27.1 and 27.2 Å [75,76] for the former, and 27.1 Å [77] for the latter. Estimations for the relative dielectric permittivity of the dipolar region of phosphatidylcholine lipid bilayers have given a wide range of values depending on the theoretical model applied [78]. The consideration of the rotating polar headgroups as an ensemble of interacting dipoles embedded in a nonhomogeneous dielectric with explicitly considering the interactions between the nearest neighborhood has derived ~30 for the headgroup dielectric constant [78], thus providing for the capacitance of the headgroup region a value of the order of 15 μF/cm². As shown by Velikonja et al. [44] at high monovalent salt concentrations the relative permittivity in the dipolar headgroup region is decreased due to a saturation effect in orientational ordering of water dipoles. Considering that our study is performed at nearly 500 times lower salt concentrations we hypothesize that the changes in the dielectric properties are much more pronounced in the hydrophobic core than in the dipolar head region. In this case, one should keep in mind that further theoretical investigations could help quantifying the sucrose effect on the dielectric permittivity of the headgroup region. Here, we suppose that the obtained alteration of ε_m is a result mainly of modulation in the hydrocarbon core of the bilayer. Hence, we deduce that in the presence of high sucrose concentrations ($\geq$200 mmol/L) the relative dielectric permittivity of hydrocarbon chains is increased by ~14% and varies between ~1.4 and 1.6 in the studied sucrose concentration range. Considering the narrower range of ε_{rm} changes upon increasing sucrose content in the bulk phase at an order of magnitude lower ionic strength than in [16] we formulate a hypothesis for a cooperative effect of sodium chloride and sucrose on membrane properties, including the bilayer specific capacitance and relative permittivity. It is supported by previous results indicating that the presence of sodium chloride modifies the effect of sucrose on the bending rigidity of lipid bilayers. We found that membrane bending elasticity measured in sucrose solutions containing 10 mM NaCl was independent of the disaccharide concentration in the aqueous surroundings [79]. This result has to be considered in the light of the ion-induced modification of sucrose-phosphatidylcholine hydrogen bond network, reported in the literature [80,81]. The capability of disaccharides to replace water molecules [51] and to create a water-like hydrogen bond network in the lipid surroundings contributes to retaining of the molecular properties of lipids [6]. Molecular dynamics simulation results have provided evidence that the hydrogen bond network of phosphatidylcholines and sucrose is partially disrupted in the presence of sodium and chloride ions [81].

5. Conclusions

The active research in the field of membrane biophysics deepens the understanding of the structural and functional membrane features in regard to vital processes in cells. In the present study a step forward is performed toward revealing the impact of small carbohydrates on the bilayer electrical properties, structure and organization. Modulation of membrane dielectric properties (capacitance and relative permittivity) is reported at moderate concentrations (above 200 mmol/L) of sucrose in the bulk phase. The above parameters are insensitive to the presence of fructose or glucose in the aqueous solution. A cooperative effect of sodium chloride and sucrose on membrane properties is suggested. As obtained by FFT EIS, all three of the small carbohydrate molecules studied here contribute to increased bilayer electric resistance. A qualitatively different behavior of the hydrocarbon fluidity in the two types of phosphatidylcholine bilayers, POPC and SOPC, upon the addition of sucrose in the aqueous surroundings is found. The hypothesis about sucrose ordering effect in membrane in the proximity of the glycerol backbone is supported here. The reduction in rotational diffusion and degree of hydration for the corresponding fluorophore is larger in the presence of the disaccharide studied. Sucrose is found to induce larger changes in lipid ordering at the glycerol level rather than in the hydrophobic core. A slight increase in the dipole potential is reported for POPC and SOPC bilayers upon the addition of sodium chloride, glucose, fructose and sucrose. The presented results are expected to be useful for the appropriate choice of carbohydrates, when the efficiency of the application targets the preservation of membrane electric properties. Gaining knowledge of the modulation of membrane molecular organization and its electric characteristics (capacitance, dielectric permittivity and resistance) by small carbohydrates would foster the elaboration of sugar-based lipid biomimetic systems for electric field-assisted applications in food industry, biotechnology and medicine.

Author Contributions: Conceptualization, V.V.; methodology, V.V., G.S. and G.P.; software, G.P.; validation, V.V., G.S. and G.P.; investigation, V.Y., O.P., A.S.-I. and K.A.; writing—original draft preparation, V.V.; writing—review and editing, G.S. and G.P.; visualization, V.V.; supervision, V.V.; project administration, V.V.; funding acquisition, V.V. and G.S. All authors have read and agreed to the published version of the manuscript.

Funding: This research and the APC were funded by Bulgarian National Science Fund, grant numbers DN18/15-2017 and DO1-154/28/08/2018, Scientific Infrastructure on Cell Technologies in Biomedicine (SICTB)—Bulgarian Ministry of Education and Science.

Institutional Review Board Statement: Not applicable.

Informed Consent Statement: Not applicable.

Data Availability Statement: Data is contained within the article.

Conflicts of Interest: The authors declare no conflict of interest. The funders had no role in the design of the study; in the collection, analyses, or interpretation of data; in the writing of the manuscript, or in the decision to publish the results.

Abbreviations

BLM	bilayer lipid membrane
Di-8-ANEPPS	4-(2-[6-(Dioctylamino)-2-naphthalenyl]ethenyl)-1-(3-sulfopropyl) pyridinium inner salt
DPH	1,6-diphenyl-1,3,5-hexatriene
FFT-EIS	fast Fourier transform electrochemical impedance spectroscopy
GP	generalized polarization
GUV	giant unilamellar vesicle
ITO	indium tin oxide
Laurdan	6-dodecanoyl-N, N-dimethyl-2-naphthylamine
LUV	large unilamellar vesicle
PC	phosphatidylcholine
PDMS	polydimethylsiloxane
POPC	1-palmitoyl-2-oleoyl-*sn*-glycero-3-phosphocholine
SOPC	1-stearoyl-2-oleoyl-*sn*-glycero-3-phosphocholine

References

1. Crowe, L.M. Lessons from nature: The role of sugars in anhydrobiosis. *Comp. Biochem. Physiol. Part A Mol. Integr. Physiol.* **2002**, *131*, 505–513. [CrossRef]
2. Panis, B.; Piette, B.; Swennen, R. Droplet vitrification of apical meristems: A cryopreservation protocol applicable to all Musaceae. *Plant Sci.* **2005**, *168*, 45–55. [CrossRef]
3. O'Brien, C.; Hiti-Bandaralage, J.; Folgado, R.; Hayward, A.; Lahmeyer, S.; Folsom, J.; Mitter, N. Cryopreservation of Woody Crops: The Avocado Case. *Plants* **2021**, *10*, 934. [CrossRef]
4. Rockinger, U.; Funk, M.; Winter, G. Current approaches of preservation of cells during (freeze-) drying. *J. Pharm. Sci.* **2021**, *110*, 2873–2893. [CrossRef]
5. Vereyken, I.J.; Chupin, V.; Demel, R.A.; Smeekens, S.C.M.; Kruijff, B.D. Fructans insert between the headgroups of phospholipids. *Biochim. Biophys. Acta Biomembr.* **2001**, *1510*, 307–320. [CrossRef]
6. Leekumjorn, S.; Sum, A.K. Molecular dynamics study on the stabilization of dehydrated lipid bilayers with glucose and trehalose. *J. Phys. Chem. B* **2008**, *112*, 10732–10740. [CrossRef]
7. van den Bogaart, G.; Hermans, N.; Krasnikov, V.; de Vries, A.H.; Poolman, B. On the decrease in lateral mobility of phospholipids by sugars. *Biophys. J.* **2007**, *92*, 1598–1605. [CrossRef]
8. Demel, R.A.; Dorrepaal, E.; Ebskamp, M.J.M.; Smeekens, J.C.M.; de Kruijff, B. Fructans interact strongly with model membranes. *Biochim. Biophys. Acta Biomembr.* **1998**, *1375*, 36–42. [CrossRef]
9. Nagle, J.F.; Jablin, M.S.; Tristram-Nagle, S. Sugar does not affect the bending and tilt moduli of simple lipid bilayers. *Chem. Phys. Lipids* **2016**, *196*, 76–80. [CrossRef]
10. Nagle, J.F.; Jablin, M.S.; Tristram-Nagle, S.; Akabori, K. What are the true values of the bending modulus of simple lipid bilayers? *Chem. Phys. Lipids* **2015**, *185*, 3–10. [CrossRef] [PubMed]
11. Vitkova, V.; Genova, J.; Mitov, M.D.; Bivas, I. Sugars in the aqueous phase change the mechanical properties of lipid mono- and bilayers. *Mol. Cryst. Liq. Cryst.* **2006**, *449*, 95–106. [CrossRef]
12. Shchelokovskyy, P.; Tristram-Nagle, S.; Dimova, R. Effect of the HIV-1 fusion peptide on the mechanical properties and leaflet coupling of lipid bilayers. *New J. Phys.* **2011**, *13*, 025004. [CrossRef] [PubMed]
13. Mitkova, D.; Vitkova, V. The aqueous surroundings alters the bending rigidity of lipid membranes. *Russ. J. Electrochem.* **2016**, *52*, 1172–1178. [CrossRef]
14. *Handbook of Electroporation*; Miklavcic, D. (Ed.) Springer: Cham, Switzerland, 2017; Volume 3, p. 2998.
15. Schwan, H.P. Dielectrophoresis and Rotation of Cells. In *Electroporation and Electrofusion in Cell Biology*; Neumann, E., Sowers, A.E., Jordan, C.A., Eds.; Springer: Boston, MA, USA, 1989. [CrossRef]
16. Vitkova, V.; Mitkova, D.; Antonova, K.; Popkirov, G.; Dimova, R. Sucrose solutions alter the electric capacitance and dielectric permittivity of lipid bilayers. *Colloids Surf. A Physicochem. Eng. Asp.* **2018**, *557*, 51–57. [CrossRef]
17. Heimburg, T. The capacitance and electromechanical coupling of lipid membranes close to transitions: The effect of electrostriction. *Biophys. J.* **2012**, *103*, 918–929. [CrossRef]
18. Huang, W.; Levitt, D.G. Theoretical calculation of the dielectric constant of a bilayer membrane. *Biophys. J.* **1977**, *17*, 111–128. [CrossRef]
19. Stern, H.A.; Feller, S.E. Calculation of the dielectric permittivity profile for a nonuniform system: Application to a lipid bilayer simulation. *J. Chem. Phys.* **2003**, *118*, 3401–3412. [CrossRef]

20. Nymeyer, H.; Zhou, H.-X. A Method to Determine Dielectric Constants in Nonhomogeneous Systems: Application to Biological Membranes. *Biophys. J.* **2008**, *94*, 1185–1193. [CrossRef]

21. *Structure and Dynamics of Membranes*; Lipowsky, R.; Sackmann, E. (Eds.) Elsevier: Amsterdam, The Netherlands, 1995.

22. *Giant Vesicles*; Luisi, P.L.; Walde, P. (Eds.) John Wiley & Sons, Ltd.: Chichester, UK, 2000.

23. *The Giant Vesicle Book*, 1st ed.; Dimova, R.; Marques, C. (Eds.) CRC Press: Boca Raton, FL, USA, 2020.

24. Allen, T.M.; Cullis, P.R. Drug delivery systems: Entering the mainstream. *Science* **2004**, *303*, 1818–1822. [CrossRef]

25. Vitkova, V.; Antonova, K.; Popkirov, G.; Mitov, M.D.; Ermakov, Y.A.; Bivas, I. Electrical resistivity of the liquid phase of vesicular suspensions prepared by different methods. *J. Phys. Conf. Ser.* **2010**, *253*, 012059. [CrossRef]

26. Khalifat, N.; Fournier, J.B.; Angelova, M.I.; Puff, N. Lipid packing variations induced by pH in cardiolipin-containing bilayers: The driving force for the cristae-like shape instability. *Biochim. Biophys. Acta Biomembr.* **2011**, *1808*, 2724–2733. [CrossRef] [PubMed]

27. Gutsmann, T.; Heimburg, T.; Keyser, U.; Mahendran, K.R.; Winterhalter, M. Protein reconstitution into freestanding planar lipid membranes for electrophysiological characterization. *Nat. Protoc.* **2015**, *10*, 188–198. [CrossRef]

28. Montal, M.; Mueller, P. Formation of Bimolecular Membranes from Lipid Monolayers and a Study of Their Electrical Properties. *Proc. Natl. Acad. Sci. USA* **1972**, *69*, 3561–3566. [CrossRef] [PubMed]

29. Blitterswijk, M.J.V.; Hoeven, R.P.V.; DerMeer, B.W.V. Lipid structural order parameters (reciprocal of fluidity) in biomembranes derived from steady-state fluorescence polarization measurements. *Biochim. Biophys. Acta Biomembr.* **1981**, *644*, 323–332. [CrossRef]

30. Parasassi, T.; Gratton, E. Membrane lipid domains and dynamics as detected by Laurdan fluorescence. *J. Fluoresc.* **1995**, *5*, 59–69. [CrossRef] [PubMed]

31. Litman, B.J.; Barenholz, Y. Fluorescent probe: Diphenylhexatriene. *Methods Enzymol.* **1982**, *81*, 678–685. [CrossRef] [PubMed]

32. Vitkova, V.; Mitkova, D.; Yordanova, V.; Pohl, P.; Bakowsky, U.; Staneva, G.; Batishchev, O. Elasticity and phase behaviour of biomimetic membrane systems containing tetraether archaeal lipids. *Colloids Surf. A Physicochem. Eng. Asp.* **2020**, *601*, 124974. [CrossRef]

33. Clarke, R.J.; Kane, D.J. Optical detection of membrane dipole potential: Avoidance of fluidity and dye-induced effects. *Biochim. Biophys. Acta Biomembr.* **1997**, *1323*, 223–239. [CrossRef]

34. Clarke, R.J. Effect of lipid structure on the dipole potential of phosphatidylcholine bilayers. *Biochim. Biophys. Acta Biomembr.* **1997**, *1327*, 269–278. [CrossRef]

35. Starke-Peterkovic, T.; Clarke, R.J. Effect of headgroup on the dipole potential of phospholipid vesicles. *Eur. Biophys. J.* **2009**, *39*, 103. [CrossRef]

36. Parisio, G.; Marini, A.; Biancardi, A.; Ferrarini, A.; Mennucci, B. Polarity-Sensitive Fluorescent Probes in Lipid Bilayers: Bridging Spectroscopic Behavior and Microenvironment Properties. *J. Phys. Chem B* **2011**, *115*, 9980–9989. [CrossRef]

37. Vlahovska, P.M.; Gracia, R.S.; Aranda-Espinoza, S.; Dimova, R. Electrohydrodynamic model of vesicle deformation in alternating electric fields. *Biophys. J.* **2009**, *96*, 4789–4803. [CrossRef]

38. Salipante, P.F.; Knorr, R.L.; Dimova, R.; Vlahovska, P.M. Electrodeformation method for measuring the capacitance of bilayer membranes. *Soft Matter* **2012**, *8*, 3810–3816. [CrossRef]

39. Dimova, R.; Bezlyepkina, N.; Jordö, M.D.; Knorr, R.L.; Riske, K.A.; Staykova, M.; Vlahovska, P.M.; Yamamoto, T.; Yang, P.; Lipowsky, R. Vesicles in electric fields: Some novel aspects of membrane behavior. *Soft Matter* **2009**, *5*, 3201–3212. [CrossRef]

40. Bockris, J.O.M.; Khan, S.U.M. The Interphasial Structure. In *Surface Electrochemistry: A Molecular Level Approach*; Springer US, Plenum Press: New York, NY, USA, 1993; pp. 59–210. [CrossRef]

41. Israelachvili, J.N. *Intermolecular and Surface Forces*; Academic Press: New York, NY, USA, 1985.

42. Gongadze, E.; Velikonja, A.; Perutkova, Š.; Kramar, P.; Maček-Lebar, A.; Kralj-Iglič, V.; Iglič, A. Ions and water molecules in an electrolyte solution in contact with charged and dipolar surfaces. *Electrochim. Acta* **2014**, *126*, 42–60. [CrossRef]

43. Iglič, A.; Gongadze, E.; Kralj-Iglič, V. Differential Capacitance of Electric Double Layer—Influence of Asymmetric Size of Ions, Thickness of Stern Layer and Orientational Ordering of Water Dipoles. *Acta Chim. Slov.* **2019**, *66*, 8. [CrossRef]

44. Velikonja, A.; Perutkova, S.; Gongadze, E.; Kramar, P.; Polak, A.; Maček-Lebar, A.; Iglič, A. Monovalent ions and water dipoles in contact with dipolar zwitterionic lipid headgroups-theory and MD simulations. *Int. J. Mol. Sci* **2013**, *14*, 2846–2861. [CrossRef] [PubMed]

45. Needham, D.; Hochmuth, R.M. Electro-mechanical permeabilization of lipid vesicles. Role of membrane tension and compressibility. *Biophys. J.* **1989**, *55*, 1001–1009. [CrossRef]

46. Popkirov, G.S.; Schindler, R.N. Validation of experimental data in electrochemical impedance spectroscopy. *Electrochim. Acta* **1993**, *38*, 861–867. [CrossRef]

47. Popkirov, G.S.; Schindler, R.N. A new approach to the problem of "good" and "bad" impedance data in electrochemical impedance spectroscopy. *Electrochim. Acta* **1994**, *39*, 2025–2030. [CrossRef]

48. Velikonja, A.; Kramar, P.; Miklavčič, D.; Maček Lebar, A. Specific electrical capacitance and voltage breakdown as a function of temperature for different planar lipid bilayers. *Bioelectrochemistry* **2016**, *112*, 132–137. [CrossRef]

49. Naumowicz, M.; Zając, M.; Kusaczuk, M.; Gál, M.; Kotyńska, J. Electrophoretic Light Scattering and Electrochemical Impedance Spectroscopy Studies of Lipid Bilayers Modified by Cinnamic Acid and Its Hydroxyl Derivatives. *Membranes* **2020**, *10*, 343. [CrossRef]

50. Batishchev, O.V.; Indenbom, A.V. Alkylated glass partition allows formation of solvent-free lipid bilayer by Montal-Mueller technique. *Bioelectrochemistry* **2008**, *74*, 22–25. [CrossRef]
51. Luzardo, M.d.C.; Amalfa, F.; Nunez, A.M.; Diaz, S.; Biondi de Lopez, A.C.; Disalvo, E.A. Effect of Trehalose and Sucrose on the Hydration and Dipole Potential of Lipid Bilayers. *Biophys. J.* **2000**, *78*, 2452–2458. [CrossRef]
52. Starke-Peterkovic, T.; Turner, N.; Else, P.L.; Clarke, R.J. Electric field strength of membrane lipids from vertebrate species: Membrane lipid composition and Na+-K+-ATPase molecular activity. *Am. J. Physiol. Regul. Integr. Comp. Physiol.* **2005**, *288*, R663–R670. [CrossRef]
53. Wang, L. Measurements and Implications of the Membrane Dipole Potential. *Annu. Rev. Biochem.* **2012**, *81*, 615–635. [CrossRef]
54. Benz, R.; Janko, K. Voltage-induced capacitance relaxation of lipid bilayer membranes Effects of membrane composition. *Biochim. Biophys. Acta* **1976**, *455*, 721–738. [CrossRef]
55. Alvarez, O.; Latorre, R. Voltage-dependent capacitance in lipid bilayers made from monolayers. *Biophys. J.* **1978**, *21*, 1–17. [CrossRef]
56. Schuster, B.; Pum, D.; Braha, O.; Bayley, H.; Sleytr, U. Self-assembled α-hemolysin pores in an S-layer-supported lipid bilayer. *Biochim. Biophys. Acta* **1998**, *1370*, 280–288. [CrossRef]
57. Gross, L.; Heron, A.; Baca, S.; Wallace, M. Determining membrane capacitance by dynamic control of droplet interface bilayer area. *Langmuir* **2011**, *27*, 14335–14342. [CrossRef] [PubMed]
58. Garten, M.; Mosgaard, L.D.; Bornschlögl, T.; Dieudonné, S.; Bassereau, P.; Toombes, G.E.S. Whole-GUV patch-clamping. *Proc. Natl. Acad Sci. USA* **2017**, *114*, 328–333. [CrossRef]
59. Mitov, M.D.; Faucon, J.F.; Méléard, P.; Bothorel, P. Thermal fluctuations of membranes. In *Advances in Supramolecular Chemistry*; Gokel, G.W., Ed.; JAI Press Inc.: Stamford, CT, USA, 1992; Volume 2, pp. 93–139.
60. Genova, J.; Vitkova, V.; Bivas, I. Registration and analysis of the shape fluctuations of nearly spherical lipid vesicles. *Phys. Rev. E* **2013**, *88*, 022707. [CrossRef] [PubMed]
61. Taylor, G.J.; Venkatesan, G.A.; Collier, C.P.; Sarles, S.A. Direct in situ measurement of specific capacitance, monolayer tension, and bilayer tension in a droplet interface bilayer. *Soft Matter* **2015**, *11*, 7592–7605. [CrossRef]
62. Beltramo, P.J.; Hooghten, R.V.; Vermant, J. Millimeter-area, free standing, phospholipid bilayers. *Soft Matter* **2016**, *12*, 4324–4331. [CrossRef]
63. Gracia, R.S.; Bezlyepkina, N.; Knorr, R.L.; Lipowsky, R.; Dimova, R. Effect of cholesterol on the rigidity of saturated and unsaturated membranes: Fluctuation and electrodeformation analysis of giant vesicles. *Soft Matter* **2010**, *6*, 1472–1482. [CrossRef]
64. Andersen, H.D.; Wang, C.; Arleth, L.; Peters, G.H.; Westh, P. Reconciliation of opposing views on membrane–sugar interactions. *Proc. Natl. Acad. Sci. USA* **2011**, *108*, 1874–1878. [CrossRef]
65. Kučerka, N.; Nieh, M.-P.; Katsaras, J. Fluid phase lipid areas and bilayer thicknesses of commonly used phosphatidylcholines as a function of temperature. *Biochim. Biophys. Acta Biomembr.* **2011**, *1808*, 2761–2771. [CrossRef]
66. Matsuki, H.; Goto, M.; Tada, K.; Tamai, N. Thermotropic and barotropic phase behavior of phosphatidylcholine bilayers. *Int. J. Mol. Sci* **2013**, *14*, 2282–2302. [CrossRef]
67. Jurkiewicz, P.; Cwiklik, L.; Jungwirth, P.; Hof, M. Lipid hydration and mobility: An interplay between fluorescence solvent relaxation experiments and molecular dynamics simulations. *Biochimie* **2012**, *94*, 26–32. [CrossRef] [PubMed]
68. Poojari, C.; Wilkosz, N.; Lira, R.B.; Dimova, R.; Jurkiewicz, P.; Petka, R.; Kepczynski, M.; Róg, T. Behavior of the DPH fluorescence probe in membranes perturbed by drugs. *Chem. Phys. Lipids* **2019**, *223*, 104784. [CrossRef]
69. Jurkiewicz, P.; Olżyńska, A.; Langner, M.; Hof, M. Headgroup Hydration and Mobility of DOTAP/DOPC Bilayers: A Fluorescence Solvent Relaxation Study. *Langmuir* **2006**, *22*, 8741–8749. [CrossRef] [PubMed]
70. Collins, K.D.; Washabaugh, M.W. The Hofmeister effect and the behaviour of water at interfaces. *Q. Rev. Biophys.* **1985**, *18*, 323–422. [CrossRef]
71. Brockman, H. Dipole potential of lipid membranes. *Chem. Phys. Lipids* **1994**, *73*, 57–79. [CrossRef]
72. Cevc, G. Hydration force and the interfacial structure of the polar surface. *J. Chem. Soc. Faraday Trans.* **1991**, *87*, 2733–2739. [CrossRef]
73. White, G.; Pencer, J.; Nickel, B.G.; Wood, J.M.; Hallett, F.R. Optical changes in unilamellar vesicles experiencing osmotic stress. *Biophys. J.* **1996**, *71*, 2701–2715. [CrossRef]
74. Nagle, J.F.; Tristram-Nagle, S. Structure of lipid bilayers. *Biochim. Biophys. Acta Biomembr.* **2000**, *1469*, 159–195. [CrossRef]
75. Tristram-Nagle, S.; Petrache, H.I.; Nagle, J.F. Structure and Interactions of Fully Hydrated Dioleoylphosphatidylcholine Bilayers. *Biophys. J.* **1998**, *75*, 917–925. [CrossRef]
76. Liu, Y.; Nagle, J.F. Diffuse scattering provides material parameters and electron density profiles of biomembranes. *Phys. Rev. E* **2004**, *69*, 040901. [CrossRef] [PubMed]
77. Kucerka, N.; Tristram-Nagle, S.; Nagle, J.F. Structure of Fully Hydrated Fluid Phase Lipid Bilayers with Monounsaturated Chains. *J. Membrane Biol.* **2005**, *208*, 193–202. [CrossRef]
78. Raudino, A.; Mauzerall, D. Dielectric properties of the polar head group region of zwitterionic lipid bilayers. *Biophys. J.* **1986**, *50*, 441–449. [CrossRef]
79. Vitkova, V.; Minetti, C.; Stoyanova-Ivanova, A. Bending rigidity of lipid bilayers in electrolyte solutions of sucrose. *Bulg. Chem. Commun.* **2020**, *52*, 35–40.

80. Valley, C.C.; Perlmutter, J.D.; Braun, A.R.; Sachs, J.N. NaCl interactions with phosphatidylcholine bilayers do not alter membrane structure but induce long-range ordering of ions and water. *J. Memb. Biol.* **2011**, *244*, 35–42. [CrossRef] [PubMed]
81. Bakarić, D.; Petrov, D.; Mouvencherya, Y.K.; Heißler, S.; Oostenbrink, C.; Schaumann, G.E. Ion-induced modification of the sucrose network and its impact on melting of freeze-dried liposomes. DSC and molecular dynamics study. *Chem. Phys. Lipids* **2018**, *210*, 38–46. [CrossRef] [PubMed]

Review

Mechanical and Electrical Interaction of Biological Membranes with Nanoparticles and Nanostructured Surfaces

Jeel Raval [1], Ekaterina Gongadze [2], Metka Benčina [3], Ita Junkar [3], Niharika Rawat [2], Luka Mesarec [2], Veronika Kralj-Iglič [4], Wojciech Góźdź [1] and Aleš Iglič [2,5,*]

1 Group of Physical Chemistry of Complex Systems, Institute of Physical Chemistry, Polish Academy of Sciences, 01-224 Warsaw, Poland; jlraval@gmail.com (J.R.); wtg@ichf.edu.pl (W.G.)
2 Laboratory of Physics, Faculty of Electrical Engineering, University of Ljubljana, 1000 Ljubljana, Slovenia; ekaterina.gongadze@fe.uni-lj.si (E.G.); niharika.rawat@fe.uni-lj.si (N.R.); mesarec.luka@gmail.com (L.M.)
3 Department of Surface Engineering and Optoelectronics, Jožef Stefan Institute, 1000 Ljubljana, Slovenia; metka.bencina@ijs.si (M.B.); ita.junkar@ijs.si (I.J.)
4 Laboratory of Clinical Biophysics, Faculty of Health Sciences, University of Ljubljana, 1000 Ljubljana, Slovenia; kraljiglic@gmail.com
5 Laboratory of Clinical Biophysics, Chair of Orthopaedics, Faculty of Medicine, University of Ljubljana, 1000 Ljubljana, Slovenia
* Correspondence: ales.iglic@fe.uni-lj.si; Tel.: +386-1-4768-825

Citation: Raval, J.; Gongadze, E.; Benčina, M.; Junkar, I.; Rawat, N.; Mesarec, L.; Kralj-Iglič, V.; Góźdź, W.; Iglič, A. Mechanical and Electrical Interaction of Biological Membranes with Nanoparticles and Nanostructured Surfaces. *Membranes* **2021**, *11*, 533. https://doi.org/10.3390/membranes11070533

Academic Editor: Monika Naumowicz

Received: 17 June 2021
Accepted: 5 July 2021
Published: 14 July 2021

Publisher's Note: MDPI stays neutral with regard to jurisdictional claims in published maps and institutional affiliations.

Abstract: In this review paper, we theoretically explain the origin of electrostatic interactions between lipid bilayers and charged solid surfaces using a statistical mechanics approach, where the orientational degree of freedom of lipid head groups and the orientational ordering of the water dipoles are considered. Within the modified Langevin Poisson–Boltzmann model of an electric double layer, we derived an analytical expression for the osmotic pressure between the planar zwitterionic lipid bilayer and charged solid planar surface. We also show that the electrostatic interaction between the zwitterionic lipid head groups of the proximal leaflet and the negatively charged solid surface is accompanied with a more perpendicular average orientation of the lipid head-groups. We further highlight the important role of the surfaces' nanostructured topography in their interactions with biological material. As an example of nanostructured surfaces, we describe the synthesis of TiO_2 nanotubular and octahedral surfaces by using the electrochemical anodization method and hydrothermal method, respectively. The physical and chemical properties of these nanostructured surfaces are described in order to elucidate the influence of the surface topography and other physical properties on the behavior of human cells adhered to TiO_2 nanostructured surfaces. In the last part of the paper, we theoretically explain the interplay of elastic and adhesive contributions to the adsorption of lipid vesicles on the solid surfaces. We show the numerically predicted shapes of adhered lipid vesicles corresponding to the minimum of the membrane free energy to describe the influence of the vesicle size, bending modulus, and adhesion strength on the adhesion of lipid vesicles on solid charged surfaces.

Keywords: lipid bilayer electrostatics; zwitterionic lipid bilayers; electric double layer; osmotic pressure; orientational degree of freedom of lipid headgroups; orientational ordering of water dipoles; adhesion of lipid vesicles; lipid bilayer elasticity; lipid vesicle shapes

1. Introduction

Biological membranes are an essential constituent of living cells. Their main role is to separate the interior of the cell from its surroundings, allowing for selective transport of specific materials across the membrane [1]. This article focuses on the interaction of biological membranes with nanostructured surfaces and nanoparticles [1–3]. The main building block of the biological membranes is the lipid bilayer with embedded inclusions such as proteins and glycolipids. Isotropic and anisotropic membrane proteins may induce local changes in the membrane curvature [1,4,5], often resulting in global changes in the cell shape [6–12]. The nonhomogeneous lateral distribution and the phase separation of membrane inclusions (nanodomains) can induce local changes in the membrane curvature and are, therefore, the driving force for transformations of the cell shape [6–9,12–14]. The biological and lipid membranes possess some degree of in-plane orientational ordering [1,7,8,10,15–18], including the nematic type of ordering [19–21], which are also important for the stability of different membrane shapes.

The configuration and shape changes of membranes are, in general, correlated with many important biological processes [1,6,13,22]. The shapes of cells are also influenced by the membrane skeleton and cytoskeleton forces [1,11,13,22–28]. Among them, the ATP consuming forces of the membrane skeleton and cytoskeleton are of major importance for sustaining different cell functions [11,12,22,28,29]. Consequently, new theoretical approaches for modeling these cell shape changes under the influence of energy-consuming active forces have been developed recently [11,12,28,29].

The focus of this paper (partially a mini review) is the interaction of nanoparticles (NPs) and nanostructured solid surfaces with cell membranes and lipid bilayers. Certain aspects of membrane–solid surface electrostatics and adhesive interactions are elucidated too.

2. Interaction of Nanoparticles with Cell Membrane

The shape and biological functions of membranes can be strongly influenced by attached, encapsulated (Figure 1) [30], or intercalated inclusions such as nanoparticles (NPs). Unique optical, electronic, catalytic, and magnetic properties of NPs make them very interesting for a variety of biomedical applications [1,31–34]. For example, functional NPs and quantum dots are potential candidates for drug delivery, as well as carriers for cancer therapy [33,35]. When NPs interact with cells, the first barrier that NPs encounter is the plasma membrane. Intra- and extracellular transport of NPs are possible by a dynamic membrane shape transformation that involves a change in the membrane curvature (encapsulation) [30,32,33,36] (Figure 1). Membrane deformations may progress passively, that is, without employing an additional energy source, driven solely by the interaction between the membrane and NPs. Viral budding comprises such an example [37]. The intracellular entry of genetic material is presently receiving considerable attention due to the COVID-19 crisis. Electrostatic interactions [38,39] may facilitate NP or virus internalization via encapsulation [30,33] (Figure 1). Understanding the interplay between the membrane elastic and electrostatic properties of the NP–membrane complex, toward the encapsulation of NPs by the cell membrane, is also relevant for cellular drug uptake, viral budding, biotechnological applications, and studying the interactions of inorganic NPs with biological membranes [32,33].

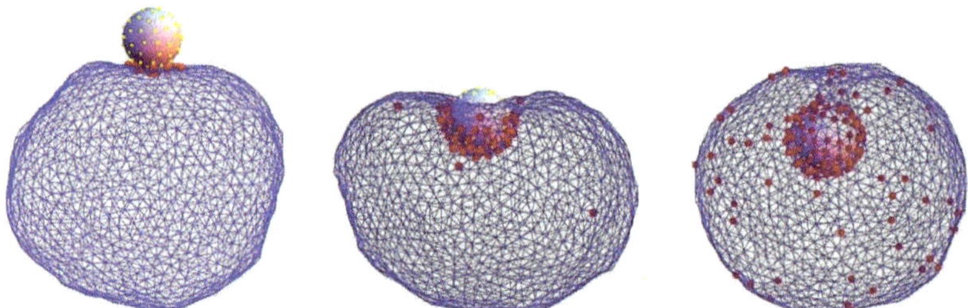

Figure 1. Encapsulation of a spherical charged particle macro-ion. Snapshots of representative configurations obtained from Monte Carlo simulations for different numbers of charged lipids (each having one unit charge) in the membrane: 15, 60, and 150 (from left to right). The right figure corresponds to the situation of nearly complete encapsulation of the macro-ion. The spherical particle carries 65 uniformly distributed, point-like cations of valence 2. Reprinted from [30] with permission of AIP Publishing.

Another possible interaction of NPs with the membrane is the attachment (adsorption) of the former on the membrane surface [34,38,40], encapsulation [30], or their intercalation in the membrane [41–43]. The resulting configuration could be driven by the NP shape, charge, size, and stiffness; it also depends on the nature of the NP–membrane interaction [30,34,38,39,41,44]. For example, hydrophobic or cationic NPs with diameters smaller than 5 nm can be successfully embedded within the membrane bilayer. On the other hand, anionic NPs of the same size or larger NPs can only interact with the outer surface of the membrane [45]. NPs interacting with membranes may induce lateral tension that results in pore formation, either transient or permanent; the pores are actually stabilized by NPs. Nanoparticles could also cluster within the membrane, and the resulting change in the membrane mechanics could significantly influence its biological function and could even result in membrane disruption [46]. Simulations demonstrated that the properties of both the membrane and the NPs are equally important in explaining the membrane uptake of the latter [30,41].

3. Interaction of Cells with Nanostructured Surfaces

Apart from exploiting the NP–membrane interactions in biomedical applications, the applications of micro- and nano-structured surfaces (Figure 2) in biomedicine have also attracted significant attention. Among them, of particular interest is the design of nanotopographic features such as biomimetic interfaces for implantable devices [47–49]. In vitro results have shown that the surface features on the nanometer scale stimulate and control several molecular and cellular events on tissue/implant interfaces, which can be observed by differences in the cell morphology, orientation, cytoskeleton organization, proliferation, and gene expression [47–49]. In the last decade, TiO_2 nanotubes fabricated via electrochemical anodization (Figures 3 and 4) have attracted significant attention toward medical applications [1,50].

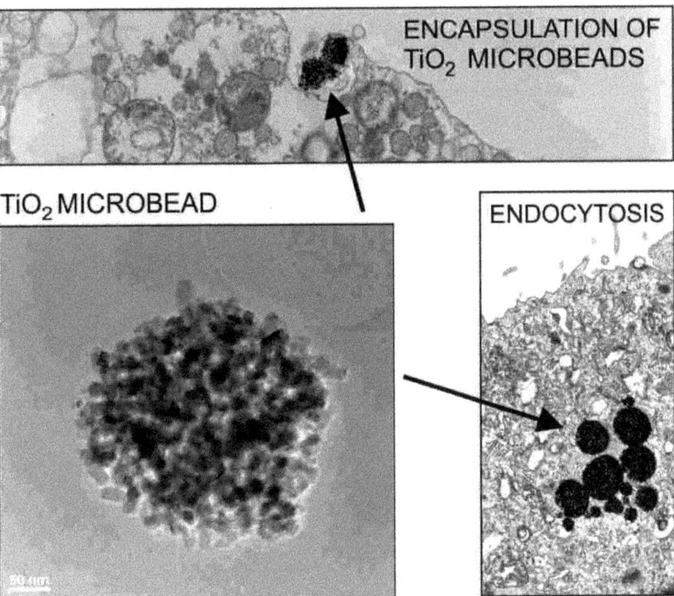

Figure 2. Encapsulation and endocytosis of TiO_2 particles (microbeads) in the cells of urinary bladder (transmission and scanning electron microscopy). Reprinted from [32,33] with the permission of AIP Publishing.

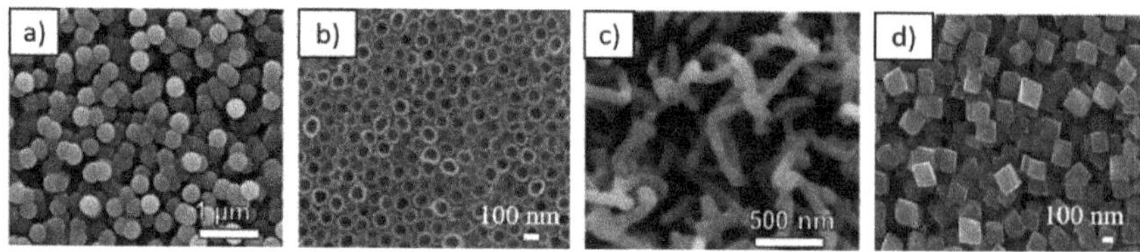

Figure 3. TiO$_2$ nanostructured surfaces: (**a**) microbeads, (**b**) nanotubes, (**c**) nanowires and (**d**) octahedral nanoparticles. Partially reprinted (**a,c**) by permission from Springer Nature: Springer Protoplasma [51] (2016).

Among others, nanotopography on TiO$_2$ surfaces exhibits reduced activation and the aggregation of platelets [52], and the controlled adhesion and proliferation of endothelial and smooth muscle cells [3]. The excellent potential of TiO$_2$ nanotubes in medicine and biotechnology is mainly due to their high effective surface area, increased surface charged density [47,48], and the possibility to vary their geometry (diameter and length), which could be specially designed/adapted for a desired biological response (cell selectivity). TiO$_2$ nanotubes have also been shown to increase selective protein adsorption [53] and, thus, they enhance the biological response. For example, studies have shown that TiO$_2$ nanotubes increase bone growth/regeneration, are antibacterial, and reduce inflammation [54,55]. Moreover, the endothelium formed on the surfaces with nanoscale topography exhibits an enhanced expression of anti-thrombogenic genes, providing a more extended coagulation cascade, probably due to a thicker oxide layer and specific topography [49].

Several reports have shown that nanotopography significantly influences cell behaviors, i.e., adhesion, proliferation, and differentiation [47,56,57]. It was recently shown that the specific nanotopography of TiO$_2$ used for cardiovascular stents can improve their bio-/hemo-compatibility via the increased adhesion and growth of human coronary artery endothelial cells and reduced adhesion and activation of platelets [3]. The surface nanostructurization of titanium (specifically the formation of TiO$_2$ nanotubes with different diameter) was shown to alter physicochemical properties, such as wettability [2] and surface chemistry, which consequently affect the interactions of TiO$_2$ nanostructured surfaces with cells [3,47,58]. In vivo studies have shown improved endothelization and reduced neointimal thickening on nanostructured stents compared to bare-metal stents [58]. Moreover, the study performed by Peng et al. [59] showed that the TiO$_2$ nanotubular surface significantly enhances endothelial cell proliferation, while, at the same time, the growth of vascular smooth muscle cells is reduced.

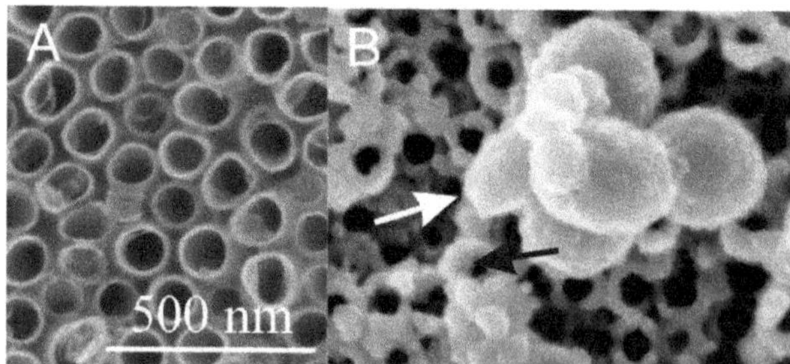

Figure 4. Scanning electron microscope (SEM) images of TiO$_2$ nanotubular surface (**A**) and the TiO$_2$ nanotubular surface that was exposed to the isolate of extracellular vesicles (white arrow) from blood (**B**). Material deposited also on the top surface of the TiO$_2$ nanotube walls is denoted by a black arrow. Adapted with permission from [60,61].

Surface nanotopography may thus strongly influence the adhesion and proliferation of cells [47]. Experimental studies of the interaction between lipid bilayers and solid surfaces have indicated that the specific structure of the surface can induce even the phase transitions in the membrane upon its adhesion to the surface [62].

Among TiO_2 nanostructured surfaces presented in Figure 3, TiO_2 nanotubular surfaces were synthesized by electrochemical anodization [3,48,53,63], while hydrothermal treatment was employed for synthesizing other structures. To illustrate the interaction of nanostructured surfaces with biological systems, two nanostructuring procedures—electrochemical anodization and hydrothermal treatment—were used for the synthesis of TiO_2 nanotubes (Figure 3b) and octahedral nanostructures (Figure 3d), respectively. In the case of electrochemical anodization, nanotubes with a diameter of 100 nm were synthesized by using HF and ethylene gylcol as electrolytes (described in detail in Refs. [60,61]), whereas for the hydrothermal method, Ti foils were exposed to a basic medium of Ti isopropoxide suspension. Briefly, in the hydrothermal method, Ti foil (0.1 mm, 99.6+%, Advent) was washed in acetone, ethanol, and water (in each for 5 min). Afterward, the samples were dried at 70 °C in the furnace. Ti isopropoxide (1 mL, 99.999% trace metals basis, Sigma Aldrich) was used as an ion precursor, and KOH (flake, 85%, Alfa Aesar) was added to the solution until the pH reached 10. The cleaned and dried Ti foil was placed at the bottom of a Teflon vessel in a hydrothermal reactor and poured with Ti isopropoxide suspension. The reaction was carried out at 200 °C for 24 h. Samples were then vigorously washed with deionized H_2O, dried under the stream of N_2, and used for further experiments.

As-prepared nanostructured surfaces were used to study interactions with blood platelets to evaluate the materials' potential thrombogenic nature as in [3]. It is of tremendous importance that metal implants, used as vascular stents, inhibit or decrease potentially fatal thrombosis and restenosis conditions. The stent surface should prevent excessive adhesion and aggregation of platelets as it can lead to blood clot formation/thrombosis. The adhesion and activation of platelets on the surface presented in Figure 5 is an indicator of how hemocompatible a material is; the lower platelet adhesion and activation, the higher is material compatibility with blood (hemocompatibility) [64,65]. It has already been recognized that protein adsorption and its conformation play an important role in platelet adhesion. Fibrinogen adsorption has been so far recognized as an important factor determining platelet adhesion and activation [66]. Studies have already shown that surface nanotopography may significantly influence platelet adhesion and activation, mainly due to nanotopography-induced changes in fibrinogen adsorption [66]. The study showed that the confirmation of adsorbed fibrinogen is highly linked to nanotopographic features of the surface. In the present study, other surface features such as wettability, roughness, and chemistry remained unchanged. However, when designing nanomaterials, it is sometimes hard to eliminate other surface features, which may even change over time. It was shown that TiO_2 nanotubular surfaces (Figure 3) tend to age with time and lose their hydrophilic characteristic [2]. The altered surface wettability and slight changes in chemistry significantly influence platelet adhesion. In this case, a decrease in the wettability of nanotubes 15, 50, and 100 nm in diameter resulted in a higher platelet adhesion and activation compared to more hydrophilic nanotubes [3]. However, the optimal surface nanostructure may significantly influence the biological response when other surface features are optimized.

In Figure 6, the scanning electron microscopy (SEM) images of plain Ti foil, TiO_2 nanotubes, and hydrothermally treated Ti foil are presented. TiO_2 nanotubes prepared by electrochemical anodization are 100 nm in diameter, while the nanostructured surface of hydrothermally treated Ti foil consists of octahedral nanoparticles with sizes of about 150–300 nm. More detailed information about surface roughness can be obtained from images taken via atomic force microscopy (AFM). The 3D image of surface topography obtained from AFM reveals that plain Ti foil has no special surface morphology, and the estimated surface roughness (Ra) evaluated from the images is about 11.7 nm. However, both nanostructured surfaces have a higher surface roughness (Ra), about 40 and 49 nm for the TiO_2 nanotubular and hydrothermal surface, respectively. It should be mentioned that in the case of nanotubes, the ability of the AFM tip to enter the hollow interior of the nanotube is limited; however, it gives us some additional information about surface topography, especially the variation in the height of nanotubes. The evaluated difference in height between nanotubes is about 190 nm. In the case of hydrothermal treatment, the difference in the height of octahedral particles is very similar, about 185 nm. Thus, the main difference between both surfaces is in the width and shape of the nano-features.

Surface chemistry should also be considered when designing the biomaterial surface. Using the XPS technique, it is possible to determine the chemical composition of the elements on the top surface (about 5 nm in depth), which interact with the biological environment. The plain, as well as the nanostructured, surfaces were analyzed by X-ray photoelectron spectroscopy (XPS). The results from XPS analysis are presented in (Table 1). An increase in titanium and oxygen concentration after both hydrothermal and anodization processes is observed compared to the control (Ti foil), partially also due to the removal of surface hydrocarbon contamination (lower carbon content). In the case of the anodization process, fluorine is also detected on the surface due to HF used as an electrolyte. Thus, after altering the surface nanotopography, changes in surface chemistry should also be considered, as they may influence biological interactions. As the wettability of surfaces is also highly correlated with biological response, it should be mentioned that both freshly synthesized nanostructured (nanotubular and octahedral) surfaces are hydrophilic with water contact angle (WCA) below 5°, while the WCA measured on plain Ti foil exhibits a more hydrophobic characteristic (WCA of about 75°). A detailed description of wettability studies can be found in Ref. [3].

Table 1. Atomic concentration (%) of detected elements by XPS on: Ti foil, hydrothermally treated Ti foil (HT), and TiO_2 nanotubular (NT) surface (diameter of the nanotubes = 100 nm).

	O	C	Ti	F
Ti foil	37.1	51.6	11.3	/
HT	47.9	31.4	20.7	/
NT	40.2	37.9	16.5	5.4

After analyzing the surface properties, the interaction of surfaces with the biological environment should be studied. For the application of biomaterial for vascular implants, the interaction with whole blood should be evaluated. According to Goodman et al. [64], platelet adhesion and activation can be evaluated from their shape. The activation of platelets can be described in three steps: adhesion, spreading/aggregation of platelets, and activation of platelets/formation of a thrombus clot [67]. The morphology of adherent platelets is commonly described as round (R), dendritic (D), spreading dendritic (SD), spreading (S), and fully spreading (FS). Platelets with F and FS are considered activated platelets (Figure 5) [64,68].

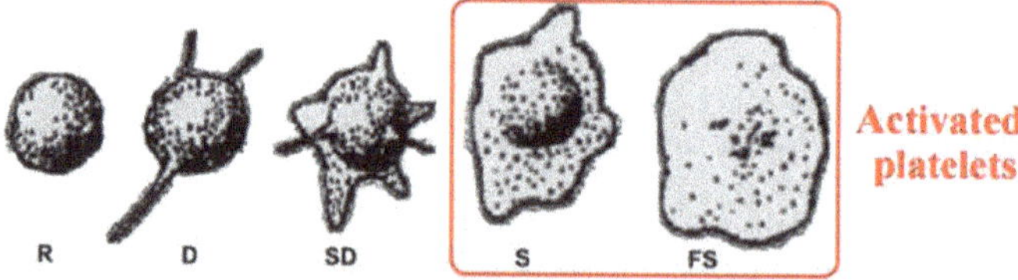

Figure 5. Platelet's morphology after the adhesion. Their shape is marked as: round (R), dendritic (D), spreading dendritic (SD), spreading (S), and fully spread (FS). Reprinted with permission from Ref. [68] copyright 2014 American Chemical Society.

The interaction of whole blood and Ti surfaces is presented in Figure 6. On the smooth, non-treated Ti foil surface (control), the platelets are preferentially in dendritic form, and platelet-to-platelet adhesion (aggregation) is observed (Figure 6g). Platelets are aggregated and activated as the specific platelet shape change is observed, i.e., the multiple filopodial extensions from the platelet body and lamellipodia formation can be seen (SD and S shape).

On the surface of the TiO$_2$ nanotubular layer, the platelets are extensively spread (S and FS shape), agglomerated, and express lamellipodia and numerous filopodia (Figure 6h). The area of spreading is significantly higher than on the surface of Ti foil.

On the surface of hydrothermally treated Ti foil, individual discoid/dendritic platelets with no obvious pseudopods are observed. Platelets are mainly in the round (R) and dendritic (D) form, indicating low surface interaction. Moreover, compared to the TiO$_2$ nanotubular layer (Figure 6h), a lower number of platelets is detected on these surfaces (Figure 6i), and platelets seem to not be in the activated form. The results presented in this section suggest that nanostructured surfaces may reduce or even increase platelet adhesion and activation. Although both nanostructured surfaces (i.e., nanotubular and octahedral) have similar surface chemistry (higher concentration of Ti and O atoms, with the addition of F for the case of the nanotubular layer) and wettability (hydrophilic) compared to plain Ti foil, their interaction with platelets seems to differ significantly in attachment and proliferation. Thus, it is important to note that a specific surface nanotopography may significantly affect the surface interaction with platelets. In Refs. [64,68], it was also shown that platelets, endothelial cells, and smooth muscle cells selectively interact with the TiO$_2$ nanotubes with various diameters, which, again, implies that surface nanotopography plays a significant role in the adhesion of biological material. Hence, by modifying the surface and with the appropriate surface chemistry, the influence of surface hydrothermally treated Ti can potentially prevent thrombosis, which could, to some extent, be correlated with its specific nanotopographic features, as the surface wettability, as well as chemistry, of both nanostructured materials was shown to be very similar.

Figure 6. SEM and AFM images of Ti foil (**a**,**d**), TiO$_2$ nanotubular surface (**b**,**e**), and TiO$_2$ octahedral surface (**c**,**f**), and the adhered blood platelets on these three kinds of surfaces (**g**–**i**).

4. On the Role of Electrostatic Interactions

Electrostatic interactions between the charged surface and the electrolyte solution result in the formation of the electric double layer (EDL) [38,69–77]. In an EDL, ions with an electric charge of the opposite sign compared to the charged surface (counterions) are accumulated close to the charged surface, and the ions with a charge of the same sign as the surface (coions) are depleted from this region [69–71,78–81]. Due to a nonhomogeneous distribution of ions in the EDL, the electric field strength is screened at larger distances from the charged surface. Due to high magnitudes of the electric field in the EDL, the water dipoles near the charged surface are strongly oriented (Figure 7) [38,82–90].

In the past, the first theoretical description of the EDL was introduced by Helmholtz [91,92], who assumed that a single layer of counterions forms at the charged surface. Later, the spatial distribution of point-like ions in the vicinity of the charged surface was described by the Boltzmann distribution function for the counter-ions and co-ions in the Poisson equation [69,70] within the so-called classical Poisson–Boltzmann (PB) model [69,70]. The PB model neglects the finite size of molecules and considers the relative permittivity as a constant throughout the whole electrolyte solution, i.e., PB approach neglects the spatial dependence of relative permittivity. A constant relative permittivity is a relatively good approximation for small magnitudes of surface charge density, but not for higher magnitudes of surface charge density where a substantial decrease in relative permittivity due to the strong orientational ordering of water dipoles in the vicinity of the charged surface was predicted [38,86,88]. In addition, the PB model also does not take into account the spatially dependent volume electric charge distribution in the zwitterionic lipids headgroup region, which depends on the interaction with neighboring charged bodies and also on the composition of the electrolyte solution [38,39,43].

The finite size of ions in the theoretical description of the EDL was first incorporated by Stern [78] with the so-called distance of closest approach and later developed further by [71,79–81]. Their work was further improved by numerous theoretical studies and simulations taking into account the asymmetry of the size of the ions, direct interactions between ions, orientational ordering of water dipoles, discrete charge distribution of the surface, quantum mechanical approach, etc. [1,72–74,76,84–86,88,90,93–115]. The physical properties of the EDL are crucial in understanding the interaction between charged surfaces in electrolyte solutions [34,38,43,116–124].

4.1. Modified Langevin Poisson–Boltzmann model

In the following, we shall describe the theoretical consideration of the electrostatic interaction (adhesion) between lipid head groups of the proximal leaflet of the lipid bilayer and charged solid surface where the orientational degree of freedom of lipid headgroups is taken into account. Among others, we shall derive, within the modified Langevin Poisson–Boltzmann [38,125,126] model, an analytical expression for the osmotic pressure between two charged surfaces, which can then also be used for the calculation of osmotic pressure between the planar lipid bilayer and charged planar surface.

We shall start with a short description of the modified Langevin Poisson–Boltzmann (LPB) model of the electric double layer [38,125,126], which presents the generalization of classic Poisson–Boltzmann (PB) theory for point-like ions by taking into account the orientational ordering of water molecules in an EDL. In the modified LPB model, the orientational ordering of water dipoles is also considered close to the saturation regime or in the saturation regime, which leads to the prediction that the relative permittivity close to the charged surface is considerably reduced. The modified LPB model also accounts for the electronic polarization of the water [38,126]. The space dependency of the relative permittivity within the modified LPB model is [38,43,126]:

$$\varepsilon_r(r) = n^2 + \frac{n_w p_0}{\varepsilon_0}\left(\frac{2+n^2}{3}\right)\left(\frac{\mathrm{L}(\gamma p_0 E(r)\beta)}{E(r)}\right),\tag{1}$$

where n is the refractive index of water, n_w is the bulk number density of water, p_0 is the magnitude of the dipole moment of a water molecule, $\mathrm{L}(u) = \coth(u) - 1/u$ is the Langevin function, $\gamma = (2+n^2)/2$, $E(r)$ is the magnitude (absolute value) of the electric field strength, $\beta = 1/kT$, and kT is the thermal energy. The above expression for the space dependency of the relative permittivity (Equation (1)) then appears in the modified LPB equation for electric potential ϕ [38,43,126]:

$$\nabla \cdot [\varepsilon_0 \varepsilon_r(r)\nabla] = 2e_0 n_0 \sinh(e_0\phi(r)\beta),\tag{2}$$

where we take into account the macroscopic (net) volume charge density of the electrolyte solution written in the form:

$$\rho(r) = e_0\, n_+(r) - e_0\, n_-(r) = -2e_0 n_0 \sinh(e_0\phi(r)\beta) \tag{3}$$

and Boltzmann distribution functions for the number densities of monovalent cations and anions:

$$n_+(r) = n_0 \exp(-e_0\phi(r)\beta), \quad n_-(r) = n_0(e_0\phi(r)\beta) \tag{4}$$

where n_0 is the bulk number density of ions. In the limit of vanishing electric field strength, the above expression for the relative permittivity (Equation (1)) yields the Onsager limit expression for bulk relative permittivity [38,43,82]:

$$\varepsilon_{r,b} = n^2 + \left(\frac{2+n^2}{3}\right)^2 \frac{n_w p_0^2 \beta}{2\varepsilon_0} \tag{5}$$

At room temperature $T = 298$ K, $p_0 = 3.1$ Debye (the Debye is 3.336×10^{-30} C/m), and $n_w/N_A = 55$ mol/L, Equation (5) gives $\varepsilon_{r,b} = 78.5$ for bulk solution. The value $p_0 = 3.1$ Debye is smaller than the corresponding value in previous similar models of electric double layers, also considering orientational ordering of the water dipole ($p_0 = 4.86$ Debye) (see, for example, [86,127]), which did not take into account the cavity field and electronic polarizability of water molecules. In addition, the model [127] predicts the increase in the relative permittivity in the direction toward the charged surface contrary to the prediction of the modified LPB model, which predicts the decrease in relative permittivity in the electrolyte solution near the charged surface [38,43,82] in agreement with experimental results. The predicted substantial increase in relative permittivity near the charged surface in [127], therefore, opposes the experimental results and defies the common principles in physics [87,123,125,126].

4.2. Osmotic Pressure between Two Charged Surfaces within Modified Langevin Poisson–Boltzmann Model

In the following, we shall derive, within the modified LPB theory, the expression for osmotic pressure between two charged planar surfaces (see Figure 7). First, we shall rearrange the modified LPB equation (Equation (2)) into a planar geometry in the form [38,43,125]:

$$-\frac{d}{dx}\left[\varepsilon_0 n^2 \frac{d\phi}{dx}\right] - n_{0w} p_0 \left(\frac{2+n^2}{3}\right) \frac{d}{dx} L(\gamma p_0 E(x)\beta) + 2e_0 n_0 \sin h(e_0\phi\beta) = 0, \tag{6}$$

where we take into account Equation (1) for relative permittivity. Equation (6) is first multiplied by $\phi' = d\phi/dx$ and then integrated to obtain [38,43]

$$-\tfrac{1}{2}\varepsilon_0 n^2 E(x)^2 + 2n_0 kT \cos h(-e_0\phi\beta) - n_w p_0\left(\tfrac{2+n^2}{3}\right) E(x)L(\gamma p_0 E(x)\beta) + \left(\tfrac{2+n^2}{3}\right)\tfrac{n_w}{\gamma\beta}\ln\left[\tfrac{\sin h(\gamma p_0 E(x)\beta)}{\gamma p_0 E(x)\beta}\right] = K, \tag{7}$$

where the constant K in Equation (7) is the local pressure between the charged surfaces. Equation (7) is equivalent to the contact theorem. In the second step, we subtract the bulk values (outside the space between the charged surfaces) from the local pressure between the charged surfaces to obtain the expression for the osmotic pressure difference $\Pi = \Pi_{inner} - \Pi_{bulk}$ in the form [38,43]:

$$\Pi = -\tfrac{1}{2}\varepsilon_0 n^2 E(x)^2 + 2n_0 kT(\cos h(-e_0\phi(x)\beta) - 1) - n_w p_0\left(\tfrac{2+n^2}{3}\right) E(x)L(\gamma p_0 E(x)\beta) + \left(\tfrac{2+n^2}{3}\right)\tfrac{n_w}{\gamma\beta}\ln\left[\tfrac{\sin h(\gamma p_0 E(x)\beta)}{\gamma p_0 E(x)\beta}\right]. \tag{8}$$

By taking into account Equation (4), we can rewrite Equation (8) in the form:

$$\Pi = -\tfrac{1}{2}\varepsilon_0 n^2 E(x)^2 + kT(n_+(x) + n_-(x) - 2n_0) - n_w p_0\left(\tfrac{2+n^2}{3}\right)E(x)L(\gamma p_0 E(x)\beta) + \left(\tfrac{2+n^2}{3}\right)\tfrac{n_w}{\gamma\beta}\ln\left[\tfrac{\sin h(\gamma p_0 E(x)\beta)}{\gamma p_0 E(x)\beta}\right].$$ (9)

The osmotic pressure is constant everywhere in the solution between the charged plates (Figure 7).

If both surfaces have equal surface charge density ($\sigma_1 = \sigma_2$), the electric field strength in the middle ($x = H/2$ in Figure 7) is zero; therefore, Equation (8) simplifies to the form [38]:

$$\Pi = 2n_0 kT(\cos h(-e_0\phi(x = H/2)\beta) - 1).$$ (10)

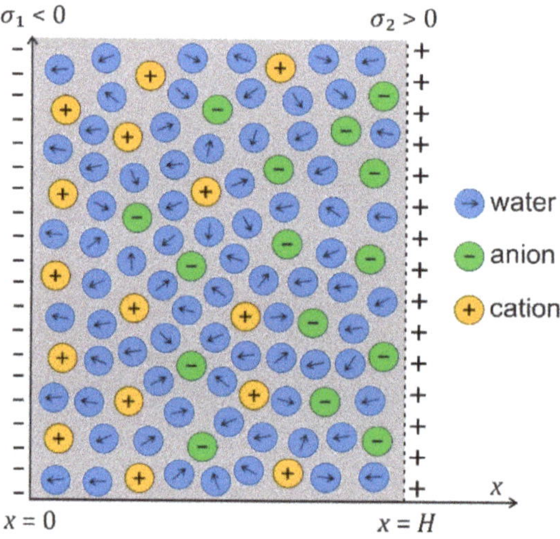

Figure 7. Schematic figure of an electrolyte solution between two charged surfaces at the distance H, where the surface charge densities $\sigma_1 < 0$ and $\sigma_2 > 0$.

For small values of $\gamma p_0 E(x)\beta$ everywhere in the solution between the two charged surfaces, we expand the third and fourth term in the above Equation (9) into series to obtain:

$$\Pi \approx -\tfrac{1}{2}\varepsilon_0\left(n^2 + \left(\tfrac{2+n^2}{3}\right)^2\tfrac{n_w p_0^2\beta}{2\varepsilon_0}\right)E(x)^2 + kT(n_+(x) + n_-(x) - 2n_0)$$
$$= -\tfrac{1}{2}\varepsilon_0\varepsilon_{r,b}E(x)^2 + kT(n_+(x) + n_-(x) - 2n_0)$$ (11)
$$= -\tfrac{1}{2}\varepsilon_0\varepsilon_{r,b}E(x)^2 + 2n_0 kT(\cos h(-e_0\phi\beta) - 1),$$

where $\varepsilon_{r,b}$ is the Onsager expression for relative permittivity, defined by Equation (5). As, in thermodynamic equilibrium, the osmotic pressure is equal everywhere between the two charged surfaces (Figure 7), we can calculate the value of the magnitude of the electric field strength in Equation (10) also at the right charged surface (Figure 7) from the corresponding boundary condition, so Equation (11) then reads

$$\Pi \approx -\frac{\sigma_2^2}{2\varepsilon_0\varepsilon_{r,b}} + 2n_0 kT(\cos h(-e_0\phi(x = H)\beta) - 1),$$ (12)

where H is the distance between the two charged surfaces.

Figure 8 presents the osmotic pressure between negatively and positively charged flat surfaces as a function of the decreasing distance (H) between them, calculated within the modified LPB model.

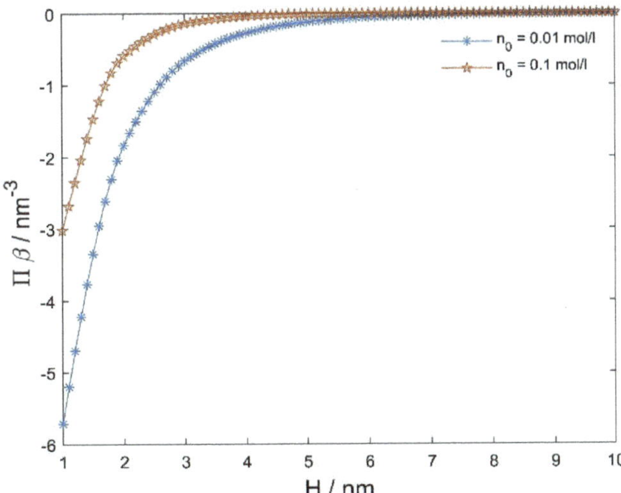

Figure 8. The calculated osmotic pressure between negatively and positively charged flat surfaces as a function of the distance between the two surfaces (H) (see Figure 7), calculated within the modified LPB model (Equations (1) and (2)) for two values of the bulk salt concentration. Other model parameters are: $\sigma_1 = 0.2$ As/m^2, $\sigma_2 = -\sigma_1$, $T = 298$ K, concentration of water $n_w / N_A = 55$ mol/L, and dipole moment of water $p_0 = 3.1$ Debye, where N_A is the Avogadro number.

4.3. Osmotic Pressure between Dipolar Zwitterionic Lipid Bilayer and Charged Rigid Surface

In the model, the zwitterionic dipolar lipid headgroup is composed of the lipids with a positively charged trimethylammonium group and a negatively charged carboxyl group, theoretically described by two charges at fixed distance, D (see Figure 9) [38,43,128]. The negative charges of the phosphate groups of dipolar (zwitterionic) lipids are described by negative surface charge density, σ_1 at $x = 0$, while the opposite charged surface with surface charge density σ_2 is located at $x = H$. The corresponding Poisson equation in a planar geometry then reads [38,43,128]:

$$\frac{d}{dx}\left[\varepsilon_0 \varepsilon_r(x)\frac{d\phi}{dx}\right] = 2e_0 n_0 \sin h(e_0\phi(x)\beta) - \rho_{ZW}(x), \tag{13}$$

where $\rho_{ZW}(x)$ is the volume charge density due to the positively charged trimethylammonium group (Figure 9):

$$\rho_{ZW}(x) = \frac{|\sigma_1|P(x)}{D} \text{ and } \rho_{ZW}(x > D) = 0, \tag{14}$$

and $P(x)$ the probability density function [38,43,128]:

$$P(x) = \Lambda\frac{\alpha \exp(-e_0\phi(x)\beta)}{\alpha \exp(-e_0\phi(x)\beta) + 1}, \quad 0 < x \leq D \tag{15}$$

The normalization constant is determined from the condition:

$$\frac{1}{D}\int_0^D P(x)dx = 1. \tag{16}$$

$P(x)$ describes the probability that the positive charge of a dipolar lipid headgroup is located at the distance x from the negatively charged surface at $x = 0$. The parameter α is equal to the ratio between the average volume of the positively charged parts of dipolar (zwitterionic) headgroups and the average volume of the salt solution in the headgroup region, meaning that the finite size of the positively charged part of the zwitterionic lipid headgroup is taken into account. The corresponding boundary conditions at $x = 0$ and $x = H$ should be taken into account [38]. The predictions of the model agree well with the results of MD simulations, as shown in [38,129].

To calculate the osmotic pressure between the zwitterionic headgroup region and positively charged surface (Figure 9), we can use Equation (8) with the input $\phi(x)$ and $E(x)$ determined from Equations (13)–(16) at appropriate boundary conditions, where the values $\phi(x)$ and $E(x)$ inserted in Equation (8) can be calculated for any $D \leq x \leq H$ because, in thermodynamic equilibrium, the osmotic pressure is equal everywhere between the two charged surfaces. However, as we are using expression Equation (8), which neglects $\rho_{ZW}(x)$ (Equation (14)), we can calculate the osmotic pressure between the dipolar zwitterionic lipid bilayer and charged rigid surface (Figure 10) by using Equation (8) only in the region $D \leq x \leq H$, where $\rho_{ZW}(x)$ is different from zero.

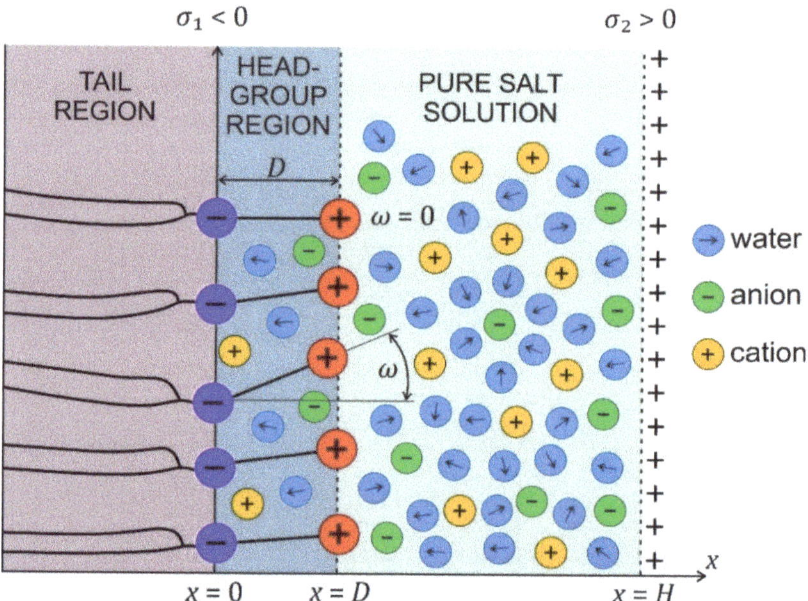

Figure 9. Schematic figure of the headgroup region composed of zwitterionic lipids with a positively charged trimethylammonium group and a negatively charged carboxyl group. The negative charges of the phosphate groups of the dipolar (zwitterionic) lipids are described by a negative surface charge density σ_1 at $x = 0$, while the electric charge due to the positively charged trimethylammonium group is described by the spatially dependent volume charge density $\rho_{ZW}(x)$, defined in the region $0 < x \leq D$ (see Equations (14)–(16)). An example of a zwitterionic lipid is SOPC.

When the zwitterionic lipid layer approaches the negatively charged surface ($\sigma_2 < 0$), the average orientation of the lipid headgroup orientation angle ($\langle \omega \rangle$) decreases with decreasing H due to the electrostatic attraction between the positively charged parts of the lipid headgroups and the negatively charged surface, as schematically shown in Figure 11, based on the results presented in [38,43]. Accordingly, the osmotic pressure between the headgroups and the negatively charged surface decreases with decreasing H, as calculated in [38,43].

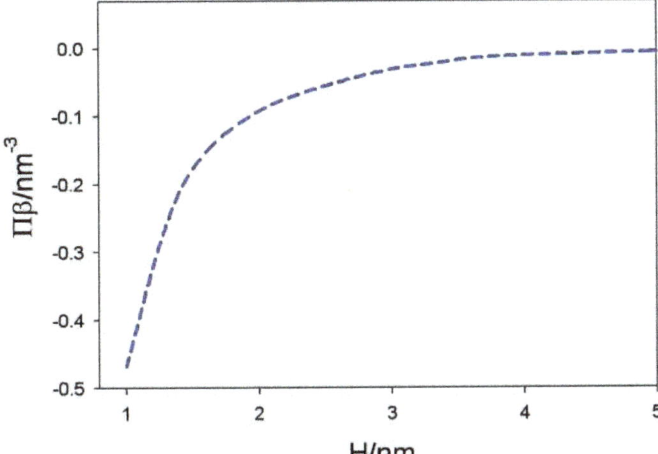

Figure 10. Calculated osmotic pressure between the dipolar headgroups and planar negatively charged surface as a function of the distance between the plane of the lipid phosphate groups and the charged surface (H) for alpha = 5. The values of model parameters are: T = 298 K, $\sigma_1 = -0.30$ As/m^2, $\sigma_2 = 0.30$ As/m^2, dipole moment of water $p_0 = 3.1$ Debye, bulk concentration of salt $n_0/N_A = 0.01$ mol/L, and concentration of water $n_w/N_A = 55$ mol/L. Reprinted from [38] with permission from Elsevier.

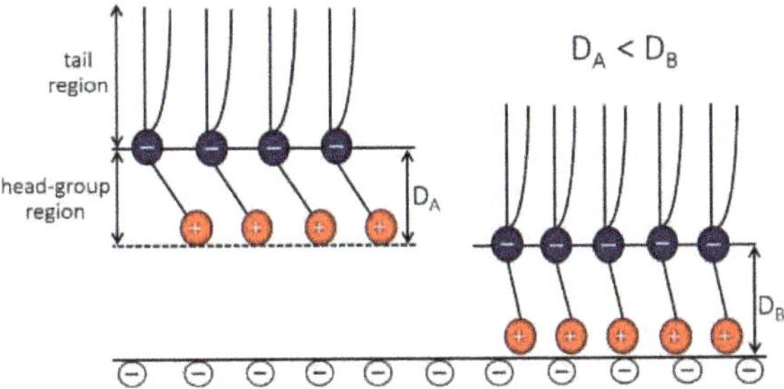

Figure 11. Schematic of the average orientation of the zwitterionic head-group at two different distances from the negatively charged surface. The figure, based on the results of theoretical modeling and MD simulations [38,128], shows that at smaller distances from the charged surface, the average orientation of the zwitterionic head-groups is more perpendicular to the charged surface.

5. Adhesion of Lipid Vesicles to Rigid Surface

In this section, we describe the interplay of membrane elasticity, geometrical constraints, and adhesive forces between the lipid bilayer and charged solid surface in the adsorption of lipid vesicles to the solid surface. The numerically calculated shapes of adhered lipid vesicles based on the system free-energy minimization are presented.

The shape of a vesicle upon adsorption to a surface is determined by the interplay of adhesion, bending, and geometrical constraints. This interplay is theoretically studied starting from a simple model in which the membrane experiences a contact potential arising from the attractive surface. Let us recall the free energy F expression of an adsorbed vesicle in terms of a simple model that takes into account the local bending energy terms, the adhesion energy and two geometrical constraints [130,131]:

$$F = \frac{1}{2}\kappa \oint (C_1 + C_2 - C_0)^2 dA + \kappa_G \oint C_1 C_2 dA - WA_c + PV + \Sigma A, \qquad (17)$$

where κ is the local bending modulus; κ_G is the Gaussian curvature modulus; C_1, C_2, and C_0 denote the two principal curvatures and the (effective) spontaneous curvature, respectively; and dA is an infinitesimal membrane area element. In the third term, W is the strength of adhesion and A_c is the contact area of the membrane and the surface. The last two terms represent the volume (V) and area (A) constraints with corresponding Lagrange multipliers P and Σ. The normalization of the membrane free energy (Equation (17)) by the bending energy of a sphere for zero spontaneous curvature $8\pi\kappa$ leads to the expression for the reduced free energy $f = F_b/8\pi\kappa$:

$$f = \frac{1}{4} \oint (c_1 + c_2 - c_0)^2 da - \frac{w}{2}\left(\frac{A_c}{A}\right) + p \oint dv + \sigma \oint da, \tag{18}$$

where $v = V/(4\pi R_s^3/3)$ is the reduced volume (see, for example, [131,132]); $a = A/4\pi R_s^2 = 1$ is the reduced area; $c_0 = C_0 R_s$, $c_1 = C_1 R_s$, and $c_2 = C_2 R_s$ are the reduced curvatures; p and σ are the reduced Lagrange multipliers; and

$$w = WR_s^2/\kappa, \tag{19}$$

is a dimensionless parameter, where $R_s = \sqrt{A/4\pi}$. The ratio $A_c/A = A_c/4\pi R_s^2$ is the reduced contact area and varies between zero for a spherical vesicle with reduced volume $v = 1.0$ and 0.5 for pancake-shaped vesicles for a very small reduced (zero) volume v. The above energy expression (Equation (18)) is minimized numerically, as described in [133]. Note that, in Figure 12 (for $c_0 = 0$), the calculated nonadhered vesicle shapes corresponding to minimal bending energy and reduced volumes $v \leq 0.591$ are stomatocytic, while the shapes for $0.592 \leq v \leq 0.651$ are oblate and, for $v \geq 0.652$, prolate (see also [134,135]). It can be seen in Figures 12 and 13 that for high reduced adhesion strength w, the calculated shapes of adhered vesicles approach the limiting shapes composed of the sections of spheres corresponding to the maximal reduced contact area at a given reduced volume v.

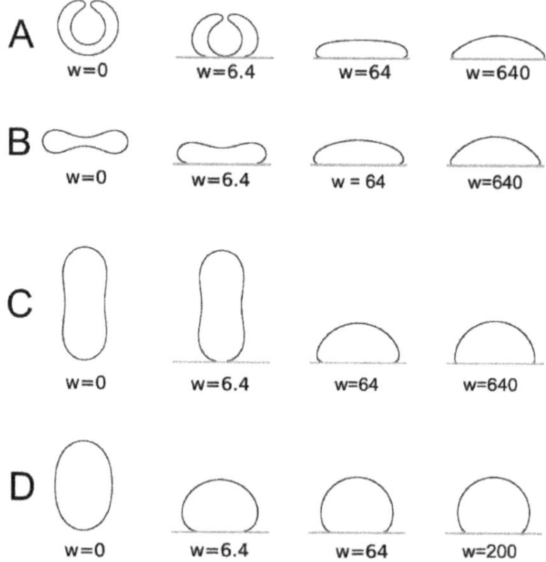

Figure 12. The calculated shapes of free (nonadsorbed) and adsorbed vesicles obtained by the minimization of the free energy given by Equation (18), determined for $c_0 = 0$ and different values of reduced volume: $v = 0.5$ (**A**), 0.6 (**B**), 0.8 (**C**), and 0.95 (**D**), and different values of reduced adhesion strength w, defined by Equation (18).

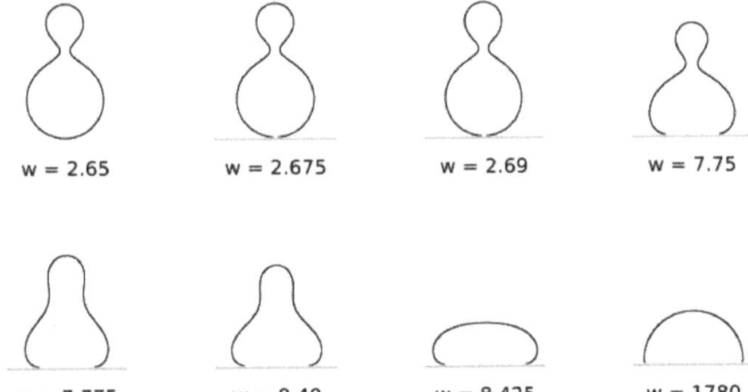

Figure 13. The calculated pear vesicle shapes of free (nonadsorbed) and adsorbed vesicles obtained by the minimization of the free energy given by Equation (18), determined for $c_0 = 2.4$, $v = 0.8$, and different values of reduced adhesion strength w, defined by Equation (19).

6. Conclusions

The growing number of nanomaterial-based commercial products (medical implants, biosensors, antibacterial surfaces, cancer therapy, etc.) has generated an increasing need for thorough scientific studies to evaluate the interactions of nanomaterials with biological cells and the influence of these interactions on the stability and growing of cells. The cell membrane is a nanoscale barrier that protects the cell and constitutes the initial contact area with the nanostructured surface. The nanomaterial–membrane interactions are strongly dependent on the membrane curvature and is influenced by geometrical/topological constraints and mechanical and electrical properties of the membranes and nanostructured surfaces [1,2,22,47,49,53,54,62,66,67]. Hence, in the future, one of the major goals of the research will be to gain a deeper understanding into the mechanisms of nanomaterial–cell membrane interactions.

In this article, we describe briefly some selected experimental methods for studying the interactions between nanostructured surfaces and biological cells. As an example, we chose TiO_2 nanotubular and octahedral surfaces and characterized them via AFM and XPS. The adhesion of human blood platelets to these surfaces was studied via SEM to elucidate the influence of the surface topography on the behavior of human cells adhered to TiO_2 nanostructured surfaces.

In the theoretical part of the article, it is shown, among others, that the electrostatic interaction between the zwitterionic lipid head groups of the proximal leaflet of the lipid bilayer and the negatively charged solid surface is accompanied with a more perpendicular average orientation of the lipid head-groups. This may induce, among others, a more tightly packed gel phase of lipids in the adhered part of the vesicle membrane and stronger orientational ordering of water dipoles between the proximal lipid layer and the supporting solid surface. In the final part of the paper, we theoretically examined the influence of the vesicle size, bending modulus, and adhesion strength on the shapes of adhered lipid vesicles. The next step would be a thorough study of the changes in the thermodynamic properties such as the phase behavior of lipid bilayers associated with the lipid bilayer response to the supported nanostructured surface [62].

Author Contributions: Conceptualization, A.I.; methodology, W.G., E.G., A.I. and V.K.-I.; software, W.G., E.G. and L.M.; numerical calculations, E.G., J.R. and L.M.; experiments, M.B., I.J. and N.R.; experiments and writing—original draft preparation, A.I., L.M., I.J. and M.B.; writing—review and editing, A.I., L.M., N.R., E.G. and V.K.-I.; visualization, L.M., E.G., I.J., M.B. and J.R.; supervision, A.I., W.G., I.J. and V.K.-I. All authors have read and agreed to the published version of the manuscript.

Funding: This project has received funding from the European Union's Horizon 2020 research and innovation programme under grant agreement no. 801338 (VES4US project) and from grant nos. P2-0232, P3-0388, J3-9262, J1-9162, and J2-8166 from the Slovenian Research Agency (ARRS). This publication is also a part of a project that has received funding from the European Union's Horizon 2020 research and innovation programme under the Marie Skłodowska-Curie grant agreement No. 711859. The scientific work was funded from the financial resources for science in the years 2017–2021 awarded by the Polish Ministry of Science and Higher Education for the implementation of an international co-financed project. WG would like to acknowledge the support from NCN grant No. 2018/30/Q/ST3/00434.

Institutional Review Board Statement: The study was conducted according to the guidelines of the Declaration of Helsinki and approved by Slovenia's Ethics Committee (number of approval; 56/03/10, date of approval; March 2010).

Informed Consent Statement: Not applicable.

Data Availability Statement: Not applicable.

Acknowledgments: We thank Marko Jeran for technical assistance.

Conflicts of Interest: The authors declare no conflict of interest.

References

1. Iglič, A.; Drobne, D.; Kralj-Iglič, V. *Nanostructures in Biological Systems: Theory and Applications*, 1st ed.; CRC Press: Boca Raton, FL, USA, 2015.
2. Kulkarni, M.; Patil-Sen, Y.; Junkar, I.; Kulkarni, C.V.; Lorenzetti, M.; Iglič, A. Wettability studies of topologically distinct titanium surfaces. *Colloids Surf. B Biointerfaces* **2015**, *129*, 47–53. [CrossRef] [PubMed]
3. Junkar, I.; Kulkarni, M.; Benčina, M.; Kovač, J.; Mrak-Poljšak, K.; Lakota, K.; Sodin-Šemrl, S.; Mozetič, M.; Iglič, A. Titanium dioxide nanotube Arrays for cardiovascular stent applications. *ACS Omega* **2020**, *5*, 7280–7289. [CrossRef] [PubMed]
4. Fošnaric, M.; Bohinc, K.; Gauger, D.R.; Iglic, A.; Kralj-Iglic, V.; May, S. The influence of anisotropic membrane inclusions on curvature elastic properties of lipid membranes. *J. Chem. Inf. Model.* **2005**, *45*, 1652–1661. [CrossRef] [PubMed]
5. Fošnarič, M.; Iglič, A.; May, S. Influence of rigid inclusions on the bending elasticity of a lipid membrane. *J. Phys. Rev. E* **2006**, *74*, 051503. [CrossRef]
6. Kralj-Iglič, V.; Pocsfalvi, G.; Mesarec, L.; Šuštar, V.; Hägerstrand, H.; Iglič, A. Minimizing isotropic and deviatoric membrane energy–An unifying formation mechanism of different cellular membrane nanovesicle types. *PLoS ONE* **2020**, *15*, e0244796. [CrossRef]
7. Kralj-Iglič, V.; Heinrich, V.; Svetina, S.; Žekš, B. Free energy of closed membrane with anisotropic inclusions. *Eur. Phys. J. B-Condens. Matter Complex Syst.* **1999**, *10*, 5–8. [CrossRef]
8. Kralj-Iglič, V.; Svetina, S.; Žekš, B. Shapes of bilayer vesicles with membrane embedded molecules. *Eur. Biophys. J.* **1996**, *24*, 311–321. [CrossRef]
9. Markin, V.S. Lateral organization of membranes and cell shapes. *Biophys. J.* **1981**, *36*, 1–19. [CrossRef]
10. Walani, N.; Torres, J.; Agrawal, A. Endocytic proteins drive vesicle growth via instability in high membrane tension environment. *Proc. Natl. Acad. Sci. USA* **2015**, *112*, E1423–E1432. [CrossRef]
11. Gov, N.S. Guided by curvature: Shaping cells by coupling curved membrane proteins and cytoskeletal forces. *Philos. Trans. R. Soc. B Biol. Sci.* **2018**, *373*, 20170115. [CrossRef]
12. Fošnarič, M.; Penič, S.; Iglič, A.; Kralj-Iglič, V.; Drab, M.; Gov, N.S. Theoretical study of vesicle shapes driven by coupling curved proteins and active cytoskeletal forces. *Soft Matter* **2019**, *15*, 5319–5330. [CrossRef]
13. Mesarec, L.; Góźdź, W.; Kralj, S.; Fošnarič, M.; Penič, S.; Kralj-Iglič, V.; Iglič, A. On the role of external force of actin filaments in the formation of tubular protrusions of closed membrane shapes with anisotropic membrane components. *Eur. Biophys. J.* **2017**, *46*, 705–718. [CrossRef]
14. Hägerstrand, H.; Mrówczyńska, L.; Salzer, U.; Prohaska, R.; Michelsen, K.A.; Kralj-Iglič, V.; Iglič, A. Curvature-dependent lateral distribution of raft markers in the human erythrocyte membrane. *Mol. Membr. Biol.* **2006**, *23*, 277–288. [CrossRef]
15. Fischer, T.M. Bending stiffness of lipid bilayers. III. Gaussian curvature. *J. Phys. II* **1992**, *2*, 337–343. [CrossRef]
16. Fischer, T.M. Bending stiffness of lipid bilayers. V. Comparison of two formulations. *J. Phys. II* **1993**, *3*, 1795–1805. [CrossRef]
17. Fournier, J.B. Nontopological saddle-splay and curvature instabilities from anisotropic membrane inclusions. *Phys. Rev. Lett.* **1996**, *76*, 4436. [CrossRef] [PubMed]
18. Safinya, C.R. Biomolecular materials: Structure, interactions and higher order self-assembly. *Colloids Surfaces A Physicochem. Eng. Asp.* **1997**, *128*, 183–195. [CrossRef]
19. Fournier, J.B.; Galatola, P. Bilayer membranes with 2D-nematic order of the surfactant polar heads. *Braz. J. Phys.* **1998**, *28*, 329–338. [CrossRef]

20. Kralj-Iglič, V.; Babnik, B.; Gauger, D.R.; May, S.; Iglič, A. Quadrupolar ordering of phospholipid molecules in narrow necks of phospholipid vesicles. *J. Stat. Phys.* **2006**, *125*, 727–752. [CrossRef]
21. Penič, S.; Mesarec, L.; Fošnarič, M.; Mrówczyńska, L.; Hägerstrand, H.; Kralj-Iglič, V.; Iglič, A. Budding and fission of membrane vesicles: A mini review. *Front. Phys.* **2020**, *8*, 342. [CrossRef]
22. Graziano, B.R.; Town, J.P.; Sitarska, E.; Nagy, T.L.; Fošnarič, M.; Penič, S.; Iglič, A.; Kralj-Iglič, V.; Gov, N.S.; Diz-Muñoz, A. Cell confinement reveals a branched-actin independent circuit for neutrophil polarity. *PLoS Biol.* **2019**, *17*, e3000457. [CrossRef]
23. Discher, D.E. Biomembrane Mechanical Properties Direct Diverse Cell Functions. In *Physics of Biological Membranes*; Springer: Berlin/Heidelberg, Germany, 2018; pp. 263–285.
24. Veksler, A.; Gov, N.S. Phase transitions of the coupled membrane-cytoskeleton modify cellular shape. *Biophys. J.* **2007**, *93*, 3798–3810. [CrossRef] [PubMed]
25. Boulbitch, A.A. Deflection of a cell membrane under application of a local force. *Phys. Rev. E* **1998**, *57*, 2123. [CrossRef]
26. Iglič, A. A possible mechanism determining the stability of spiculated red blood cells. *J. Biomech.* **1997**, *30*, 35–40. [CrossRef]
27. Iglič, A.; Kralj-Iglic, V.; Hägerstrand, H. Amphiphile induced echinocyte-spheroechinocyte transformation of red blood cell shape. *Eur. Biophys. J.* **1998**, *27*, 335–339. [CrossRef] [PubMed]
28. Alimohamadi, H.; Smith, A.S.; Nowak, R.B.; Fowler, V.M.; Rangamani, P. Non-uniform distribution of myosin-mediated forces governs red blood cell membrane curvature through tension modulation. *PLoS Comput. Biol.* **2020**, *16*, e1007890. [CrossRef]
29. Penič, S.; Fošnarič, M.; Mesarec, L.; Iglič, A.; Kralj-Iglič, V. Active forces of myosin motors may control endovesiculation of red blood cells. *Acta Chim. Slov.* **2020**, *67*, 674–681. [CrossRef]
30. Fošnarič, M.; Iglič, A.; Kroll, D.M.; May, S. Monte Carlo simulations of complex formation between a mixed fluid vesicle and a charged colloid. *J. Chem. Phys.* **2009**, *131*, 09B610. [CrossRef]
31. Rossi, G.; Monticelli, L. Gold nanoparticles in model biological membranes: A computational perspective. *Biochim. Biophys. Acta (BBA) Biomembr.* **2016**, *1858*, 2380–2389. [CrossRef]
32. Imani, R.; Veranič, P.; Iglič, A.; Kreft, M.E.; Pazoki, M.; Hudoklin, S. Combined cytotoxic effect of UV-irradiation and TiO_2 microbeads in normal urothelial cells, low-grade and high-grade urothelial cancer cells. *Photochem. Photobiol. Sci.* **2015**, *14*, 583–590. [CrossRef]
33. Imani, R.; Dillert, R.; Bahnemann, D.W.; Pazoki, M.; Apih, T.; Kononenko, V.; Repar, N.; Kralj-Iglič, V.; Boschloo, G.; Drobne, D. Multifunctional gadolinium-Doped mesoporous TiO_2 nanobeads: Photoluminescence, enhanced spin relaxation, and reactive oxygen species photogeneration, beneficial for cancer diagnosis and treatment. *Small* **2017**, *13*, 1700349. [CrossRef] [PubMed]
34. Goršak, T.; Drab, M.; Križaj, D.; Jeran, M.; Genova, J.; Kralj, S.; Lisjak, D.; Kralj-Iglič, V.; Iglič, A.; Makovec, D. Magneto-mechanical actuation of barium-hexaferrite nanoplatelets for the disruption of phospholipid membranes. *J. Colloid Interface Sci.* **2020**, *579*, 508–519. [CrossRef] [PubMed]
35. Wang, D.; Sun, Y.; Liu, Y.; Meng, F.; Lee, R.J. Clinical translation of immunoliposomes for cancer therapy: Recent perspectives. *Expert Opin. Drug Deliv.* **2018**, *15*, 893–903. [CrossRef] [PubMed]
36. Bahrami, A.H.; Lipowsky, R.; Weikl, T.R. The role of membrane curvature for the wrapping of nanoparticles. *Soft Matter* **2016**, *12*, 581–587. [CrossRef]
37. Cooper, A.; Paran, N.; Shaul, Y. The earliest steps in hepatitis B virus infection. *Biochim. Biophys. Acta (BBA) Biomembr.* **2003**, *1614*, 89–96. [CrossRef]
38. Gongadze, E.; Velikonja, A.; Perutkova, Š.; Kramar, P.; Maček-Lebar, A.; Kralj-Iglič, V.; Iglič, A. Ions and water molecules in an electrolyte solution in contact with charged and dipolar surfaces. *Electrochim. Acta* **2014**, *126*, 42–60. [CrossRef]
39. Santhosh, P.B.; Velikonja, A.; Perutkova, Š.; Gongadze, E.; Kulkarni, M.; Genova, J.; Eleršič, K.; Iglič, A.; Kralj-Iglič, V.; Ulrih, N.P. Influence of nanoparticle–membrane electrostatic interactions on membrane fluidity and bending elasticity. *Chem. Phys. Lipids* **2014**, *178*, 52–62. [CrossRef]
40. Šarić, A.; Cacciuto, A. Self-assembly of nanoparticles adsorbed on fluid and elastic membranes. *Soft Matter* **2013**, *9*, 6677–6695. [CrossRef]
41. Daniel, M.; Řezníčková, J.; Handl, M.; Iglič, A.; Kralj-Iglič, V. Clustering and separation of hydrophobic nanoparticles in lipid bilayer explained by membrane mechanics. *Sci. Rep.* **2018**, *8*, 1–7. [CrossRef]
42. Yi, X.; Gao, H. Incorporation of soft particles into lipid vesicles: Effects of particle size and elasticity. *Langmuir* **2016**, *32*, 13252–13260. [CrossRef]
43. Velikonja, A.; Santhosh, P.B.; Gongadze, E.; Kulkarni, M.; Eleršič, K.; Perutkova, Š.; Kralj-Iglič, V.; Ulrih, N.P.; Iglič, A. Interaction between dipolar lipid headgroups and charged nanoparticles mediated by water dipoles and ions. *Int. J. Mol. Sci.* **2013**, *14*, 15312–15329. [CrossRef] [PubMed]
44. Wi, H.S.; Lee, K.; Pak, H.K. Interfacial energy consideration in the organization of a quantum dot–lipid mixed system. *J. Phys. Condens. Matter* **2008**, *20*, 494211.
45. Contini, C.; Hindley, J.W.; Macdonald, T.J.; Barritt, J.D.; Ces, O.; Quirke, N. Size dependency of gold nanoparticles interacting with model membranes. *Commun. Chem.* **2020**, *3*, 1–12. [CrossRef] [PubMed]
46. Liu, Y.; Zhang, Z.; Zhang, Q.; Baker, G.L.; Worden, R.M. Biomembrane disruption by silica-core nanoparticles: Effect of surface functional group measured using a tethered bilayer lipid membrane. *Biochim. Biophys. Acta (BBA) Biomembr.* **2014**, *1838*, 429–437. [CrossRef]

47. Gongadze, E.; Kabaso, D.; Bauer, S.; Slivnik, T.; Schmuki, P.; van Rienen, U.; Iglič, A. Adhesion of osteoblasts to a nanorough titanium implant surface. *Int. J. Nanomed.* **2011**, *6*, 1801.

48. Kulkarni, M.; Mazare, A.; Gongadze, E.; Perutkova, Š.; Kralj-Iglič, V.; Milošev, I.; Schmuki, P.; Iglič, A.; Mozetič, M. Titanium nanostructures for biomedical applications. *Nanotechnology* **2015**, *26*, 062002. [CrossRef]

49. Mohan, C.C.; Chennazhi, K.P.; Menon, D. In vitro hemocompatibility and vascular endothelial cell functionality on titania nanostructures under static and dynamic conditions for improved coronary stenting applications. *Acta Biomater.* **2013**, *9*, 9568–9577. [CrossRef]

50. Benčina, M.; Iglič, A.; Mozetič, M.; Junkar, I. Crystallized TiO_2 nanosurfaces in biomedical applications. *Nanomaterials* **2020**, *10*, 1121. [CrossRef]

51. Imani, R.; Pazoki, M.; Zupančič, D.; Kreft, M.E.; Kralj-Iglič, V.; Veranič, P.; Iglič, A. Biocompatibility of different nanostructured TiO_2 scaffolds and their potential for urologic applications. *Protoplasma* **2016**, *253*, 1439–1447. [CrossRef]

52. Benčina, M.; Junkar, I.; Mavrič, T.; Iglič, A.; Kralj-Iglic, V.; Valant, M. Performance of annealed TiO_2 nanotubes in interactions with blood platelets. *Mater. Tehnol.* **2019**, *53*, 791–795. [CrossRef]

53. Kulkarni, M.; Mazare, A.; Park, J.; Gongadze, E.; Killian, M.S.; Kralj, S.; von der Mark, K.; Iglič, A.; Schmuki, P. Protein interactions with layers of TiO_2 nanotube and nanopore arrays: Morphology and surface charge influence. *Acta Biomater.* **2016**, *45*, 357–366. [CrossRef] [PubMed]

54. Ding, X.; Zhou, L.; Wang, J.; Zhao, Q.; Lin, X.; Gao, Y.; Li, S.; Wu, J.; Rong, M.; Guo, Z. The effects of hierarchical micro/nanosurfaces decorated with TiO_2 nanotubes on the bioactivity of titanium implants in vitro and in vivo. *Int. J. Nanomed.* **2015**, *10*, 6955.

55. Neacsu, P.; Mazare, A.; Schmuki, P.; Cimpean, A. Attenuation of the macrophage inflammatory activity by TiO_2 nanotubes via inhibition of MAPK and NF-κB pathways. *Int. J. Nanomed.* **2015**, *10*, 6455.

56. Yang, L.; Gong, Z.; Lin, Y.; Chinthapenta, V.; Li, Q.; Webster, T.J.; Sheldon, B.W. Disordered topography mediates filopodial extension and morphology of cells on stiff materials. *Adv. Funct. Mater.* **2017**, *27*, 1702689. [CrossRef]

57. Guilak, F.; Cohen, D.M.; Estes, B.T.; Gimble, J.M.; Liedtke, W.; Chen, C.S. Control of stem cell fate by physical interactions with the extracellular matrix. *Cell Stem Cell* **2009**, *5*, 17–26. [CrossRef]

58. Cherian, A.M.; Joseph, J.; Nair, M.B.; Nair, S.V.; Maniyal, V.; Menon, D. Successful reduction of neointimal hyperplasia on stainless steel coronary stents by titania nanotexturing. *ACS Omega* **2020**, *5*, 17582–17591. [CrossRef]

59. Peng, L.; Eltgroth, M.L.; LaTempa, T.J.; Grimes, C.A.; Desai, T.A. The effect of TiO_2 nanotubes on endothelial function and smooth muscle proliferation. *Biomaterials* **2009**, *30*, 1268–1272. [CrossRef]

60. Kralj-Iglič, V.; Dahmane, R.; Bulc, T.G.; Trebše, P.; Battelino, S.; Kralj, M.B.; Benčina, M.; Bohinc, K.; Božič, D.; Debeljak, M. From extracellular vesicles to global environment: A cosmopolite SARS-CoV-2 virus. *Int. J. Clin. Stud. Med.Case Rep.* **2020**, *4*, 1–23.

61. Kulkarni, M.; Šepitka, J.; Junkar, I.; Benčina, M.; Rawat, N.; Mazare, A.; Rode, C.; Gokhale, S.; Schmuki, P.; Daniel, M.; et al. Mechanical properties of anodic titanium dioxide nanostructures. *Mater. Technol.* **2021**, *55*, 19–24. [CrossRef]

62. Bibissidis, N.; Betlem, K.; Cordoyiannis, G.; Prista-von Bonhorst, F.; Goole, J.; Raval, J.; Daniel, M.; Góźdź, W.; Iglič, A.; Losada-Pérez, P. Correlation between adhesion strength and phase behaviour in solid-supported lipid membranes. *J. Mol. Liq.* **2020**, *320*, 114492. [CrossRef]

63. Benčina, M.; Junkar, I.; Zaplotnik, R.; Valant, M.; Iglič, A.; Mozetič, M. Plasma-induced crystallization of TiO_2 nanotubes. *Materials* **2019**, *12*, 626. [CrossRef]

64. Goodman, S.L.; Grasel, T.G.; Cooper, S.L.; Albrecht, R.M. Platelet shape change and cytoskeletal reorganization on polyurethaneureas. *J. Biomed. Mater. Res.* **1989**, *23*, 105–123. [CrossRef]

65. Schettini, N.; Jaroszeski, M.J.; West, L.; Saddow, S.E. Hemocompatibility assessment of 3C-SiC for cardiovascular applications. In *Silicon Carbide Biotechnology*; Elsevier: Waltham, MA, USA, 2012; pp. 153–208.

66. Firkowska-Boden, I.; Helbing, C.; Dauben, T.J.; Pieper, M.; Jandt, K.D. How Nanotopography-induced conformational changes of fibrinogen affect platelet adhesion and activation. *Langmuir* **2020**, *36*, 11573–11580. [CrossRef]

67. Zarka, R.; Horev, M.B.; Volberg, T.; Neubauer, S.; Kessler, H.; Spatz, J.P.; Geiger, B. Differential modulation of platelet adhesion and spreading by adhesive ligand density. *Nano Lett.* **2019**, *19*, 1418–1427. [CrossRef]

68. Ding, Y.; Yang, Z.; Bi, C.W.C.; Yang, M.; Xu, S.L.; Lu, X.; Huang, N.; Huang, P.; Leng, Y. Directing vascular cell selectivity and hemocompatibility on patterned platforms featuring variable topographic geometry and size. *ACS Appl. Mater. Interfaces* **2014**, *6*, 12062–12070. [CrossRef]

69. Gouy, M. Sur la constitution de la charge électrique à la surface d'un électrolyte. *J. Phys. Le Radium* **1910**, *9*, 457–468. [CrossRef]

70. Chapman, D.L.L. A contribution to the theory of electrocapillarity. *London Edinburgh Dublin Philos. Mag. J. Sci.* **1913**, *25*, 475–481. [CrossRef]

71. Freise, V. Zur theorie der diffusen doppelschicht. *Z. Elektrochem. Ber. Bunsenges. Phys. Chem.* **1952**, *56*, 822–827.

72. Torrie, G.M.; Valleau, J.P. Electrical double layers. I. Monte Carlo study of a uniformly charged surface. *J. Chem. Phys.* **1980**, *73*, 5807–5816. [CrossRef]

73. Kenkel, S.W.; Macdonald, J.R. A lattice model for the electrical double layer using finite-length dipoles. *J. Chem. Phys.* **1984**, *81*, 3215–3221. [CrossRef]

74. Outhwaite, C.W.; Bhuiyan, L.B. A modified Poisson–Boltzmann equation in electric double layer theory for a primitive model electrolyte with size-asymmetric ions. *J. Chem. Phys.* **1986**, *84*, 3461–3471. [CrossRef]

75. McLaughlin, S. The electrostatic properties of membranes. *Ann. Rev. Biophys. Biophys. Chem.* **1989**, *18*, 113–136. [CrossRef]
76. Kralj-Iglič, V.; Iglič, A. A simple statistical mechanical approach to the free energy of the electric double layer including the excluded volume effect. *J. Phys. II* **1996**, *6*, 477–491. [CrossRef]
77. Bivas, I. Electrostatic and mechanical properties of a flat lipid bilayer containing ionic lipids: Possibility for formation of domains with different surface charges. *Colloids Surfaces A Physicochem. Eng. Asp.* **2006**, *282*, 423–434. [CrossRef]
78. Stern, O. Zur Theorie der elektrolytischen Doppelschicht. *Z. Elektrochem.* **1924**, *30*, 508–516.
79. Bikerman, J.J. Structure and capacity of electrical double layer. *Philos. Mag.* **1942**, *33*, 384–397. [CrossRef]
80. Wicke, E.; Eigen, M. Über den Einfluß des Raumbedarfs von Ionen in wäßriger Lösung auf ihre Verteilung in elektrischen Feld und ihre Aktivitätskoeffizienten. *Z. Elektrochem.* **1952**, *56*, 551–561.
81. Eigen, M.; Wicke, E. The thermodynamics of electrolytes at higher concentration. *J. Phys. Chem.* **1954**, *58*, 702–714. [CrossRef]
82. Onsager, L. Electric moments of molecules in liquids. *J. Am. Chem. Soc.* **1936**, *58*, 1486–1493. [CrossRef]
83. Booth, F. The dielectric constant of water and the saturation effect. *J. Chem. Phys.* **1951**, *19*, 391–394. [CrossRef]
84. Outhwaite, C.W. A treatment of solvent effects in the potential theory of electrolyte solutions. *Mol. Phys.* **1976**, *31*, 1345–1357. [CrossRef]
85. Outhwaite, C.W. Towards a mean electrostatic potential treatment of an ion-dipole mixture or a dipolar system next to a plane wall. *Mol. Phys.* **1983**, *48*, 599–614. [CrossRef]
86. Iglič, A.; Gongadze, E.; Bohinc, K. Excluded volume effect and orientational ordering near charged surface in solution of ions and Langevin dipoles. *Bioelectrochemistry* **2010**, *79*, 223–227. [CrossRef]
87. Bazant, M.Z.; Kilic, M.S.; Storey, B.D.; Ajdari, A. Towards an understanding of induced-charge electrokinetics at large applied voltages in concentrated solutions. *Adv. Colloid Interface Sci.* **2009**, *152*, 48–88. [CrossRef] [PubMed]
88. Gongadze, E.; Iglič, A. Asymmetric size of ions and orientational ordering of water dipoles in electric double layer model-an analytical mean-field approach. *Electrochim. Acta* **2015**, *178*, 541–545. [CrossRef]
89. Gongadze, E.; Mesarec, L.; Kralj-Iglic, V.; Iglic, A. Asymmetric finite size of ions and orientational ordering of water in electric double layer theory within lattice model. *Mini Rev. Med. Chem.* **2018**, *18*, 1559–1566. [CrossRef]
90. Iglič, A.; Gongadze, E.; Kralj-Iglič, V. Differential Capacitance of Electric Double Layer-Influence of Asymmetric Size of Ions, Thickness of Stern Layer and Orientational Ordering of Water Dipoles. *Acta Chim. Slov.* **2019**, *66*, 534–541. [CrossRef]
91. Helmholtz, H. Ueber einige Gesetze der Vertheilung elektrischer Ströme in körperlichen Leitern, mit Anwendung auf die thierisch-elektrischen Versuche. *Ann. Phys.* **1853**, *165*, 353–377. [CrossRef]
92. Helmholtz, H.V. Studien über electrische Grenzschichten. *Ann. Phys.* **1879**, *243*, 337–382. [CrossRef]
93. Liu, C.; Elvati, P.; Majumder, S.; Wang, Y.; Liu, A.P.; Violi, A. Predicting the time of entry of nanoparticles in lipid membranes. *ACS Nano* **2019**, *13*, 10221–10232. [CrossRef] [PubMed]
94. Torrie, G.M.; Valleau, J.P. Electrical double layers. 4. Limitations of the Gouy-Chapman theory. *J. Phys. Chem.* **1982**, *86*, 3251–3257. [CrossRef]
95. Nielaba, P.; Forstmann, F. Packing of ions near an electrolyte-electrode interface in the hnc/lmsa approximation to the rpm model. *Chem. Phys. Lett.* **1985**, *117*, 46–48. [CrossRef]
96. Plischke, M.; Henderson, D. Pair correlation functions and density profiles in the primitive model of the electric double layer. *J. Chem. Phys.* **1988**, *88*, 2712–2718. [CrossRef]
97. Kornyshev, A.A. Double-layer in ionic liquids: Paradigm change? *J. Phys. Chem. B* **2007**, *111*, 5545–5557. [CrossRef]
98. Mier-y-Teran, L.; Suh, S.H.; White, H.S.; Davis, H.T. A nonlocal free-energy density-functional approximation for the electrical double layer. *J. Chem. Phys.* **1990**, *92*, 5087–5098. [CrossRef]
99. Strating, P.; Wiegel, F.W. Effects of excluded volume on the electrolyte distribution around a charged sphere. *J. Phys. A Math. Gen.* **1993**, *26*, 3383. [CrossRef]
100. Lee, J.W.; Nilson, R.H.; Templeton, J.A.; Griffiths, S.K.; Kung, A.; Wong, B.M. Comparison of molecular dynamics with classical density functional and poisson–boltzmann theories of the electric double layer in nanochannels. *J. Chem. Theory Comput.* **2012**, *8*, 2012–2022. [CrossRef]
101. Quiroga, M.A.; Xue, K.-H.; Nguyen, T.-K.; Tułodziecki, M.; Huang, H.; Franco, A.A. A multiscale model of electrochemical double layers in energy conversion and storage devices. *J. Electrochem. Soc.* **2014**, *161*, E3302. [CrossRef]
102. Bandopadhyay, A.; Shaik, V.A.; Chakraborty, S. Effects of finite ionic size and solvent polarization on the dynamics of electrolytes probed through harmonic disturbances. *Phys. Rev. E* **2015**, *91*, 042307. [CrossRef] [PubMed]
103. Lian, C.; Liu, K.; van Aken, K.L.; Gogotsi, Y.; Wesolowski, D.J.; Liu, H.L.; Jiang, D.E.; Wu, J.Z. Enhancing the capacitive performance of electric double-layer capacitors with ionic liquid mixtures. *ACS Energy Lett.* **2016**, *1*, 21–26. [CrossRef]
104. Drab, M.; Kralj-Iglič, V. Diffuse electric double layer in planar nanostructures due to Fermi-Dirac statistics. *Electrochim. Acta* **2016**, *204*, 154–159. [CrossRef]
105. Budkov, Y.A.; Kolesnikov, A.L.; Goodwin, Z.A.H.; Kiselev, M.G.; Kornyshev, A.A. Theory of electrosorption of water from ionic liquids. *Electrochim. Acta* **2018**, *284*, 346–354. [CrossRef]
106. Budkov, Y.A. Nonlocal statistical field theory of dipolar particles in electrolyte solutions. *J. Phys. Condens. Matter* **2018**, *30*, 344001. [CrossRef] [PubMed]
107. Dubtsov, A.V.; Pasechnik, S.V.; Shmeliova, D.V.; Saidgaziev, A.S.; Gongadze, E.; Iglič, A.; Kralj, S. Liquid crystalline droplets in aqueous environments: Electrostatic effects. *Soft Matter* **2018**, *14*, 9619–9630. [CrossRef]

108. Gavish, N.; Elad, D.; Yochelis, A. From solvent-free to dilute electrolytes: Essential components for a continuum theory. *J. Phys. Chem. Lett.* **2018**, *9*, 36–42. [CrossRef] [PubMed]

109. Gavish, N. Poisson-Nernst-Planck equations with steric effects—Non-convexity and multiple stationary solutions. *Phys. D Nonlinear Phenom.* **2018**, *368*, 50–65. [CrossRef]

110. Kruczek, J.; Chiu, S.-W.; Varma, S.; Jakobsson, E.; Pandit, S.A. Interactions of monovalent and divalent cations at palmitoyl-oleoyl-phosphatidylcholine interface. *Langmuir* **2019**, *35*, 10522–10532. [CrossRef]

111. Liu, X.; Tian, R.; Ding, W.; Wu, L.; Li, H. Role of ionic polarization and dielectric decrement in the estimation of surface potential of clay particles. *Eur. J. Soil Sci.* **2019**, *70*, 1073–1081. [CrossRef]

112. May, S. Differential capacitance of the electric double layer: Mean-field modeling approaches. *Curr. Opin. Electrochem.* **2019**, *13*, 125–131. [CrossRef]

113. Cruz, C.; Kondrat, S.; Lomba, E.; Ciach, A. Effect of proximity to ionic liquid-solvent demixing on electrical double layers. *J. Mol. Liq.* **2019**, *294*, 111368. [CrossRef]

114. Guardiani, C.; Gibby, W.; Barabash, M.; Luchinsky, D.; Khovanov, I.; McClintock, P. Prehistory probability distribution of ionic transitions through a graphene nanopore. In Proceedings of the 25th International Conference on Noise and Fluctuations—ICNF 2019, Neuchatel, Switzerland, 18–21 June 2019.

115. Khademi, M.; Barz, D.P.J. Structure of the Electrical Double Layer Revisited: Electrode Capacitance in Aqueous Solutions. *Langmuir* **2020**, *36*, 4250–4260. [CrossRef]

116. Kjellander, R.; Marčelja, S. Interaction of charged surfaces in electrolyte solutions. *Chem. Phys. Lett.* **1986**, *127*, 402–407. [CrossRef]

117. Evans, D.F.; Wennerström, H. *The Colloidal Domain: Where Physics, Chemistry, Biology, and Technology Meet*; Wiley-VCH: Weinheim, Germany, 1999.

118. Butt, H.J.; Graf, K.; Kappl, M. *Physics and Chemistry of Interfaces*; Wiley-VCH: Weinheim, Germany, 2003.

119. Bohinc, K.; Iglič, A.; May, S. Interaction between macroions mediated by divalent rod-like ions. *Europhys. Lett.* **2004**, *68*, 494. [CrossRef]

120. Urbanija, J.; Bohinc, K.; Bellen, A.; Maset, S.; Iglič, A.; Kralj-Iglič, V.; Sunil Kumar, P.B. Attraction between negatively charged surfaces mediated by spherical counterions with quadrupolar charge distribution. *J. Chem. Phys.* **2008**, *129*, 105101. [CrossRef]

121. Perutková, Š.; Frank, M.; Bohinc, K.; Bobojevič, G.; Zelko, J.; Rozman, B.; Kralj-Iglič, V.; Iglič, A. Interaction between equally charged membrane surfaces mediated by positively and negatively charged macro-ions. *J. Membr. Biol.* **2010**, *236*, 43–53. [CrossRef]

122. Israelachvili, J.N. *Intermolecular and Surface Forces*, 3rd ed.; Academic Press: London, UK, 2011.

123. Prasanna Misra, R.; Das, S.; Mitra, S.K. Electric double layer force between charged surfaces: Effect of solvent polarization. *J. Chem. Phys.* **2013**, *138*, 114703. [CrossRef]

124. Gimsa, J.; Wysotzki, P.; Perutkova, S.; Weihe, T.; Elter, P.; Marszałek, P.; Kralj-Iglič, V.; Müller, T.; Iglič, A. Spermidine-induced attraction of like-charged surfaces is correlated with the pH-dependent spermidine charge: Force spectroscopy characterization. *Langmuir* **2018**, *34*, 2725–2733. [CrossRef]

125. Gongadze, E.; van Rienen, U.; Kralj-Iglič, V.; Iglič, A. Langevin Poisson-Boltzmann equation: Point-like ions and water dipoles near a charged surface. *Gen. Physiol. Biophys.* **2011**, *30*, 130–137. [CrossRef]

126. Drab, M.; Gongadze, E.; Kralj-Iglič, V.; Iglič, A. Electric Double Layer and Orientational Ordering of Water Dipoles in Narrow Channels within a Modified Langevin Poisson-Boltzmann Model. *Entropy* **2020**, *22*, 1054. [CrossRef] [PubMed]

127. Abrashkin, A.; Andelman, D.; Orland, H. Dipolar Poisson-Boltzmann equation: Ions and dipoles close to charge interfaces. *Phys. Rev. Lett.* **2007**, *99*, 077801. [CrossRef]

128. Velikonja, A.; Perutkova, Š.; Gongadze, E.; Kramar, P.; Polak, A.; Maček-Lebar, A.; Iglič, A. Monovalent ions and water dipoles in contact with dipolar zwitterionic lipid headgroups-theory and MD simulations. *Int. J. Mol. Sci.* **2013**, *14*, 2846–2861. [CrossRef]

129. Lebar, A.M.; Velikonja, A.; Kramar, P.; Iglič, A. Internal configuration and electric potential in planar negatively charged lipid head group region in contact with ionic solution. *Bioelectrochemistry* **2016**, *111*, 49–56. [CrossRef]

130. Lipowsky, R.; Seifert, U. Adhesion of vesicles and membranes. *Mol. Cryst. Liq. Cryst.* **1991**, *202*, 17–25. [CrossRef]

131. Deuling, H.J.; Helfrich, W. The curvature elasticity of fluid membranes: A catalogue of vesicle shapes. *J. Phys.* **1976**, *37*, 1335–1345. [CrossRef]

132. Iglič, A.; Kralj-Iglič, V.; Majhenc, J. Cylindrical shapes of closed lipid bilayer structures correspond to an extreme area difference between the two monolayers of the bilayer. *J. Biomech.* **1999**, *32*, 1343–1347. [CrossRef]

133. Góźdź, W.T. Spontaneous curvature induced shape transformations of tubular polymersomes. *Langmuir* **2004**, *20*, 7385–7391. [CrossRef] [PubMed]

134. Seifert, U.; Berndl, K.; Lipowsky, R. Shape transformations of vesicles: Phase diagram for spontaneous-curvature and bilayer-coupling models. *Phys. Rev. A* **1991**, *44*, 1182. [CrossRef] [PubMed]

135. Mesarec, L.; Góźdź, W.; Iglič, A.; Kralj-Iglič, V.; Virga, E.G.; Kralj, S. Normal red blood cells' shape stabilized by membrane's in-plane ordering. *Sci. Rep.* **2019**, *9*, 1–11. [CrossRef]

 *membranes*

Article

The Effect of Submicron Polystyrene on the Electrokinetic Potential of Cell Membranes of Red Blood Cells and Platelets

Marcin Zając [1], Joanna Kotyńska [2], Mateusz Worobiczuk [2], Joanna Breczko [2] and Monika Naumowicz [2,*]

[1] Doctoral School of Exact and Natural Sciences, University of Bialystok, K. Ciolkowskiego 1K, 15-245 Bialystok, Poland; m.zajac@uwb.edu.pl

[2] Department of Physical Chemistry, Faculty of Chemistry, University of Bialystok, K. Ciolkowskiego 1K, 15-245 Bialystok, Poland; joannak@uwb.edu.pl (J.K.); mworobiczuk@gmail.com (M.W.); j.luszczyn@uwb.edu.pl (J.B.)

* Correspondence: monikan@uwb.edu.pl; Tel.: +48-8573-880-71

Abstract: In recent years, many scientists have studied the effects of polymer micro- and nanostructures on living organisms. As it turns out, plastic can be a component of the blood of livestock, eaten by humans around the globe. Thus, it seems important to investigate possible changes in the physico-chemical parameters and morphology of the cell membranes of blood morphotic elements (red blood cells and platelets) under the influence of polymer particles. The article presents research in which cell membranes were exposed to plain polystyrene (PS) and amino-functionalized polystyrene (PS-NH$_2$) of two different sizes. The polymers were characterized by infrared spectroscopy and dynamic light-scattering methods. To analyze possible changes caused by polymer exposure in the structure of the membranes, their zeta potentials were measured using the electrophoretic light-scattering technique. The concentration of the polymers, as well as the exposure time, were also taken into the consideration during the research. Based on the obtained results, we concluded that 100 and 200 nm PS, as well as 100 nm PS-NH$_2$, internalize into the cells. On the contrary, 200 nm PS-NH$_2$ particles attach to cell membranes. Our study clearly shows that particle size and surface chemistry determine the interaction with biological membranes.

Keywords: polymers; erythrocytes; platelets; electrophoretic light scattering; dynamic light scattering; FTIR spectroscopy

Citation: Zając, M.; Kotyńska, J.; Worobiczuk, M.; Breczko, J.; Naumowicz, M. The Effect of Submicron Polystyrene on the Electrokinetic Potential of Cell Membranes of Red Blood Cells and Platelets. *Membranes* **2022**, *12*, 366. https://doi.org/10.3390/membranes12040366

Academic Editor: Francisco Monroy

Received: 1 March 2022
Accepted: 22 March 2022
Published: 26 March 2022

Publisher's Note: MDPI stays neutral with regard to jurisdictional claims in published maps and institutional affiliations.

1. Introduction

Nowadays, it is difficult to imagine functioning without polymers. Polymers are used in various industries, such as construction [1], electrical engineering [2], and medicine [3]. In recent years, more and more researchers from around the world have become interested in the characteristics of nano- and microplastic. Submicron polymer particles are defined as molecules of plastic smaller than 5 mm in diameter [4]. They are separated into two types: (1) Primary plastics, which enter the environment in their micro- and nanoscopic state, and (2) secondary plastics, which are the result of continuous environmental impacts on plastic litter and yield progressively smaller polymer fragments. Polystyrene (PS) is one of the most commonly used polymers in the production of plastic materials, together with polyethylene and polypropylene [5]. Its use is found mainly in food packaging, which, due to COVID-19, became a very busy branch of industry. Synthetic polymers have become a serious concern for the life and health safety of living organisms, as it is becoming clearer just how omnipresent plastic is in the ecosystem. Human activity is the main cause of contamination of the environment with polymers. Additionally, plastic production is continually increasing. It is estimated to reach 33 billion tons in 2050 [6]. The consequences of polymer particle debris for wildlife are becoming better and better documented.

Factors affecting the breakdown of polymers into smaller structures include UV radiation [7] or the influence of microorganisms [8–10]. The presence of polymers has already been evidenced in soil [11], freshwater [12], oceans [13], plants [14], animal organisms [15], and food products [16]. Mass production of plastic articles in society poses great risks, as these items should be properly disposed of or recycled. If these processes do not take place, polymers will certainly end up in the environment. Given plastic's ubiquitous nature and small dimensions, its ingestion and subsequent impact on living organisms is a growing cause for concern. At cellular and molecular levels, alterations of immunological responses, neurotoxic effects, and the onset of genotoxicity have been observed in water-dwelling organisms exposed to polystyrene particles [17,18].

Domestic pigs (*Sus domestica*) are particularly useful in biomedical research for several reasons. This species is characterized by high fertility and occurs in many genotypes and phenotypes, which is important for the selection of appropriate genetic material for medicinal studies [19–22]. The usage of biological material from pigs raises less opposition in laboratory tests than tests on dogs or cats. *Sus domestica* is very important in medical sciences such as surgery [23,24] and pharmacy [25,26] because of correlations between their bodies and humans. Pigs are not only comparable in size, but also have analogies to humans in the structure of their digestive, nervous, circulatory, and urinary systems. These similarities in the anatomy and physiology of *Sus domestica* and *Homo sapiens* are one reason biological material from pigs can be used for reliable modeling of cellular processes of human bodies.

Soil is also believed to be a pathway for micropolymers to enter living organisms. As it was found out, plastic particles can enter plants grown in heavily polluted areas [27]. This could lead to the consumption of polymers while eating contaminated vegetables by humans or animals. Other studies have shown that digesting processes do not break down plastic, and synthetic polymers can be a component of manure [28]. Scientists discovered polymers in the muscle tissue of marine animals, such as turtles [29] or fish [30]. It is indicated that plastic particles can also cross the blood–brain barrier. The consequence of this is the accumulation of polymers in the microglia cells, worsening their ability to multiply and leading to their death [31]. This information sheds a different light on the effects of polymers on human cells, including cell membranes.

Thus, food consumption can be a direct path of entry to the human body for micropolymers. Scientists believe that polymer molecules can move up the food chain thanks to their presence in the bloodstream of animals [32]. Thus, any type of meat could be a potential route for micro- and nanopolymers to enter the human system. This could have very serious consequences for health and wellness. Many tests have to be carried out to assess the toxicity and transport pathway of micropolymers from livestock, e.g., pigs, to the human body.

Experimental studies of the electrical properties of natural cell membranes can contribute to our knowledge of their properties, function, and structure. The zeta potential (otherwise known as the electrokinetic potential, ζ) of membranes is an extremely important electrical parameter that depends on the composition of the cytoplasmic membrane. This parameter defines the stability of a system and allows one to determine whether modifications of the natural system with polymers affect changes in the structure of the natural membrane. The zeta potential varies according to many parameters, such as temperature, pH, conductivity (ionic strength), and solvent (viscosity). Thus, small modifications to any of these parameters may have a significant influence on the zeta potential values [33]. One technique used to estimate zeta potential is electrophoretic light scattering (ELS), a technique based on dynamic light scattering (DLS), in which the shift in the frequency or oscillation phase of the laser beam depends on the mobility of particles/cells in an alternating electric field [34].

The purpose of the research presented in this article is to assess the influence of PS particles on the zeta potential of *Sus domestica* blood components—red blood cells (RBCs), also referred to as erythrocytes and platelets (thrombocytes). We chose polystyrene as it does not generate reactive oxygen species in the presence of cells, and membrane damage by oxidative stress is unlikely [35]. Two sizes of two types of polystyrene were used: 100 nm plain polystyrene (PS-100) and 200 nm plain polystyrene (PS-200), as well as 100 nm polystyrene with amino groups (PS-NH$_2$-100) and 200 nm polystyrene with amino groups (PS-NH$_2$-200). In addition, different particle concentrations and exposure times to the polymer were applied. Based on the knowledge of the authors, such a thorough analysis has never been performed before.

2. Materials and Methods

2.1. Materials

2.1.1. Domestic Pig Blood

The blood was obtained from the Bost Meat Plant in Turosn Koscielna from adult pigs of the species *Sus domestica*. Animal blood was collected directly into sterile test tubes made of unbreakable plastic with buffered sodium citrate in a ratio of 1:9 to biological material. Those test tubes were delivered to the laboratory, where individual morphotic elements of blood were immediately isolated.

2.1.2. Polymers

The following polymers were used to modify the biological membranes of the morphotic blood elements of *Sus domestica*: PS-100 and PS-200 (purchased from Sigma Aldrich, Saint Louis, MO, USA), PS-NH$_2$-100 (purchased from Polysciences, Hirschberg an der Bergstrasse, Germany), and PS-NH$_2$-200 (purchased from Bang Laboratories, Fishers, IN, USA). Each polymer was used at the concentrations of 0.002, 0.01, 0.1, and 0.5 mg/mL. The polymers were diluted to the appropriate concentration using 0.155 M sodium chloride, in which the blood was also suspended for physiological conditions. In preliminary experimental studies, the stability and durability of the tested polymers in the electrolyte solution were investigated using the DLS method. The tests were performed immediately after mixing morphotic blood elements with polymers and after 1 h and 3 h of exposure to plastic particles.

2.2. Methods

2.2.1. Isolation of Erythrocytes from Blood

The blood samples (2 mL volume) were centrifuged at 900 rpm for 8 min at room temperature. The red blood cells fell to the bottom of the tube, while the thrombocyte-rich plasma was removed from the system and further isolation was performed. The erythrocytes were washed three times with a 0.155 M NaCl solution at 3000 rpm for 15 min to obtain a clean erythrocyte-rich system. Finally, the washed red blood cells were suspended in an electrolyte solution (0.155 M NaCl) to prepare them for measurement via electrophoretic light scattering.

2.2.2. Isolation of Thrombocytes from Plasma

The plasma was centrifuged at 4000 rpm for 8 min. Thrombocytes remained at the bottom of the tube after centrifugation. The plasma was removed from the vial. The platelets were washed three times with 0.155 M sodium chloride, as were the erythrocytes. Finally, the thrombocytes were added to the electrolyte solution (0.155 M NaCl). This solution was then used during measurements performed by the ELS technique.

2.2.3. Comparison of Different Polymers by FTIR Spectroscopy

The Fourier transform infrared (FTIR) measurements were performed on the Nicolet 6700 spectrometer (Thermo Scientific, Madison, WI, USA) in transmission mode using the KBr method. First, the stock solutions of the different PS microspheres (PS-100, PS-200, PS-NH$_2$-100, and PS-NH$_2$-200) were diluted with distilled water to a concentration of 0.1 mg/mL. Next, aqueous PS solutions were frozen with liquid nitrogen and lyophilized in a freeze dryer (Christ Alpha 1–2 LD plus with double–chamber, Osterode am Harz, Germany) for 24 h under 0.013 mbar pressure. The obtained powder samples were mixed and ground in an agate mortar with dry KBr at a mass ratio of 1:100. The powders pressed into KBr pellets were subjected to FTIR analysis. The FTIR spectra of different PS microspheres were collected at room temperature in the range 4000–400 cm^{-1} with a wavenumber resolution of 4 cm^{-1}.

2.2.4. Determination of Polymers Size

The DLS method was used to determine the durability and stability of polymer particles in 0.155 M sodium chloride at pH = 7.4 at 25 °C. The polymer particles suspended in the electrolyte were illuminated with a helium-neon (He–Ne) laser with a wavelength of 633 nm. Measurements were performed with a Zetasizer Nano ZS apparatus (Malvern Instruments, Malvern, UK). Using the Stokes–Einstein equation, the apparatus converted the particle velocity caused by Brownian motion into particle distribution. When determining the particle size, the Non-Invasive Back Scatter method was used, in which the detector is positioned at an angle of 173 degrees. The measured value is the hydrodynamic diameter. Size distributions by numbers were obtained. The resulting values were burdened with an error expressed as a standard deviation. All PS had a polydispersity index (called PDI, used to describe the variation in size) lower than 0.3. Additionally, analysis using the ELS technique was carried out that allowed for determining the zeta potential values of the tested systems.

2.2.5. Determination of Zeta Potential

The zeta potential of both erythrocytes and thrombocytes was obtained by performing microelectrophoretic assessments on samples using the ELS technique. The experiment was performed as a pH function using a WTW InoLab pH 3310 laboratory meter (WTW, Weinheim, Germany). Blood components were suspended in a 0.155 M NaCl solution and titrated to the desired pH (range 3–12) with strong acid (HCl) and strong base (NaOH) solutions, prepared with sodium chloride to maintain the strength of the ionic solution constant. Polystyrene was added to NaCl to achieve the final concentrations of each polymer type: 0.002, 0.01, 0.1, and 0.5 mg/mL. The particles suspended in sodium chloride were exposed to an electric field during the measurements. The Zetasizer Nano ZS apparatus (Malvern, Great Britain) uses the Phase Analysis Light-Scattering technique to determine the electrophoretic mobility of the tested particles. Then, the electrophoretic mobility is converted into zeta potential. Experiments were conducted eight times with similar results obtained.

2.2.6. Statistical Analysis

The data obtained in this study are expressed as mean $\pm$ SD. The data were analyzed by use of standard statistical analyses, namely one-way ANOVA with Scheffe's F test for multiple comparisons to determine the significance between different groups. Values of $p < 0.05$ were considered significant.

3. Results and Discussion

An investigation into the characteristics of the polymers used in the research was carried out to analyze the occurrence of possible flaws in the commercial product that could affect the conducted analysis. The hydrodynamic size and stability of the submicron polystyrene particles were measured using the DLS method. FTIR spectroscopy was also utilized during the research for identity analysis.

To investigate the interactions between the cell membranes of morphotic blood components (erythrocytes and thrombocytes) and various polystyrene particles (PS-100, PS-200, PS-NH$_2$-100, PS-NH$_2$-200), several experimental studies were carried out using the ELS method. These interactions could potentially change the values of the electrokinetic potential of the systems studied. Both the influence of the polymer concentration and the exposure time of the cell membrane to the polymer were taken into the account during the experiment. Each of the measurements was performed as a function of the H$^+$ ion concentration.

3.1. Characteristics of Polymers Used in the Study

Structural studies of PS particles differing in size and/or the presence of amine groups on their surface were carried out using FTIR spectroscopy. Firstly, each of the analyzed polymers was suspended in water, and then the obtained sample was frozen and freeze-dried. The FTIR analysis of PS-100, PS-200, PS-NH$_2$-100, and PS-NH$_2$-200 is shown in Figure 1.

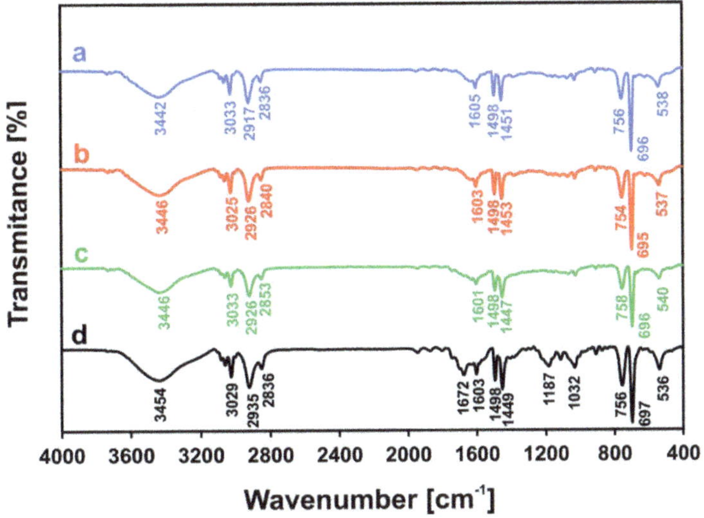

Figure 1. FTIR spectra of (**a**) PS–100, (**b**) PS–200, (**c**) PS–NH$_2$–100, and (**d**) PS–NH$_2$–200.

As expected, the recorded spectra of all samples showed bands confirming the structure of polystyrene [36–38]. The bands at approximately 3025–3033 cm^{-1} were attributed to the C–H stretching vibration of the benzene ring, while the signals at approximately 2917–2935 and 2836–2853 were assigned to the aliphatic –CH$_2$– stretching vibration. Further proof of the presence of the benzene ring in the structure of the analyzed particles is the bands at 1605, 1498, and 1451 cm^{-1} (e.g., values for PS-100, Figure 1a), corresponding to the stretching vibrations of the aromatic C=C bonds. The bands at approximately 756 and 696 cm^{-1} (e.g., values for PS-100, Figure 1a) were characteristic of the aromatic substitution pattern [36,37] and were assigned to the C–H out-of-plane bending vibration. A broad peak at approximately 3442–3454 cm^{-1}, corresponding to the stretching vibrations of O–H, indicated the existence of hydroxyl groups likely coming from the water. All obtained spectra (Figure 1a–d) looked much the same, differing slightly in the position of the bands that are characteristic of polystyrene. Additionally, the size of the analyzed particles did not result in any significant structural differences (PS-100, PS-200, Figure 1a,b). However, in the case of the amino-functionalized PS particles (PS-NH$_2$-100, PS-NH$_2$-200, Figure 1c,d) additional bands were observed, thus confirming their surface modification. The signal observed in the PS-NH$_2$-200 spectrum (Figure 1d) at 1672 cm^{-1} and the bands appearing at 1187 and 1032 cm^{-1} were assigned to the N-H bending vibrations characteristic of primary amines and the stretching vibrations of C-N bonds, respectively. The analogous peaks in the FTIR spectrum recorded for PS-NH$_2$-100 (Figure 1c) exhibited a low intensity and are not clearly visible. The reason for this is likely the smaller size of the tested submicron particles and the proportionally smaller number of –NH$_2$ groups on its surface.

Parameters (size and zeta potential) characterizing the polymers were measured in 0.155 M sodium chloride at pH = 7.4 with the Zetasizer Nano ZS apparatus. The results obtained using the DLS technique are shown in Figure 2.

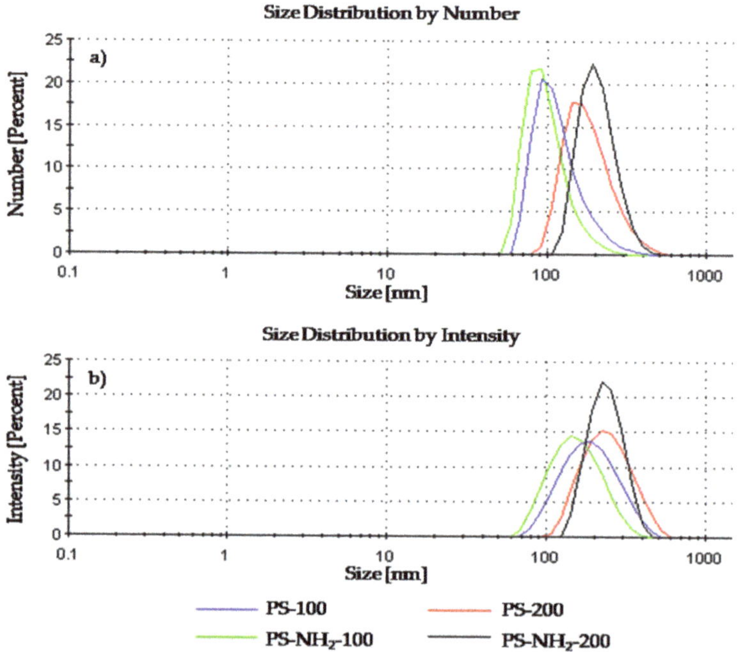

Figure 2. Polystyrene size distribution by (**a**) number, (**b**) intensity.

The plots of the polymer size distribution by number and intensity indicated the formation of monomodal fractions. The data obtained from dependences of the particles' diameter by number range from 110 to 207 nm, and regarding intensity, range from 160 nm to 249 nm. The values registered for PS-100 and PS-NH$_2$-100 differ slightly from the ones confirmed by the companies, while for PS-200 and PS-NH$_2$-200, the measurements are consistent with the information provided by the producer.

The presence of other fractions, above 3000 nm in diameter, should not be ruled out. All size data obtained, including the standard deviation (SD) and polydispersity index (PDI) values, are collated in Table 1.

Table 1. The parameters characterizing the polymers (0.155 M NaCl, pH = 7.4).

Polymer	Size by Number [nm]	SD	Size by Intensity [nm]	SD	PDI	Zeta Potential [mV]	SD
PS-100	119.90	47.04	197.40	76.24	0.257	−31.00	0.99
PS-200	185.70	67.86	249.40	86.00	0.205	−46.10	2.11
PS-NH$_2$-100	110.10	34.01	160.70	58.50	0.231	−24.90	1.20
PS-NH$_2$-200	207.90	54.82	238.40	57.04	0.055	−28.60	1.32

Summaries for the values of zeta potential are also compiled in Table 1, measured for each polymer using the ELS technique. The lowest zeta potential values were recorded for plain polystyrene particles with a size of 200 nm (−46.10 ± 2.11 mV). The remaining polymers (PS-100, PS-NH$_2$-100, and PS-NH$_2$-200) show ζ values ranging from −24.90 ± 1.20 mV to −31.00 ± 0.99 mV.

Generally, a large positive or negative value of the zeta potential (lower than −30 mV and higher than +30 mV) indicates physical stability due to the electrostatic repulsion of individual particles. On the other hand, a small value of the electrokinetic potential may cause the aggregation or flocculation of particles due to the van der Waals forces [39]. Based on data collected in Table 1, it can be concluded that all polymers are characterized by good stability.

3.2. The Effect of Polystyrene Polymers on the Zeta Potential of Morphotic Components of Pig Blood

3.2.1. The Effect of Polystyrene Polymers Concentration

The ELS technique was used to provide insight into the possible changes in zeta potential values, due to treating erythrocytes and thrombocytes' cell membranes with PS-100, PS-200, PS-NH$_2$-100, or PS-NH$_2$-200. In order to obtain the pH-dependent ζ value, the systems, suspended in a 0.155 M NaCl solution, were titrated to the appropriate pH with concentrated NaOH or HCl.

Figure 3 shows representative plots of the electrokinetic potential vs. pH obtained for cell membranes of red blood cells modified with submicron polystyrene particles differing in size and/or the presence of amino groups on their surface. As it can be observed, an increase in the positive value of the zeta potential was observed alongside a decrease in the pH value. Conversely, as the pH increased, the negative values of ζ increased until they reached a plateau.

Upon analyzing data depicted in Figure 3, it can be noted that mixing red blood cells with three types of polystyrene particles (PS-100, PS-200, and PS-NH$_2$-100, Figure 3a–c) did not cause statistically significant changes in ζ values compared to the control sample, which was pure erythrocytes suspended in 0.155 M sodium chloride. These changes were not noticeable in almost the entire range of polymer concentrations, with exceptions at extreme pH values where destruction of the membrane structure occurred. This was seen for membranes measured in the electrolyte solution containing 0.01 mg/mL PS-200 at pH~3 or 0.1 mg/mL PS-100 at pH~12. The lack of zeta potential changes allowed us to conclude that PS-100, PS-200, and PS-NH$_2$-100 were internalized by the cell. The entry mechanisms for nanoparticles into cells are still not yet understood. None of the endocytic pathways, all of which involve vesicle formation in an actin-mediated process, are likely to account for nanoparticle translocation. The translocation of particles may also occur via non-specific pathways, including diffusion, trans-membrane channels, electrostatic, hydration, van der Waals forces, or steric interactions [40]. Moreover, PS particles within cells are not membrane-bound and hence have direct access to intracellular proteins, organelles, and DNA, which may greatly enhance their toxic potential [41].

In contrast, statistically significant changes in the ζ values of the erythrocyte cell membranes were observed after their treatment with PS-NH$_2$-200 in the entire pH range studied, as shown in Figure 3d and Table 2.

Table 2. The zeta potential of the erythrocytes' cell membranes after exposure to PS-NH$_2$-200 (C = 0.002, 0.01, 0.1, and 0.5 mg/mL).

pH	Control	0.002	0.01	0.1	0.5
3	6.07 ± 0.19	6.86 ± 0.98	9.97 ± 0.65 [a,b]	11.68 ± 0.86 [a,b,c]	13.72 ± 0.90 [a,b,c,d]
4	−4.44 ± 0.50	−0.02 ± 1.65 [a]	6.85 ± 0.67 [a,b]	10.35 ± 0.67 [a,b,c]	3.23 ± 0.75 [a,b,c,d]
5	−7.15 ± 1.28	−2.37 ± 0.52 [a]	0.00 ± 0.83 [a,b]	0.86 ± 1.00 [a,b]	−3.78 ± 1.31 [a,c,d]
6	−8.20 ± 0.51	−7.03 ± 0.45	−8.53 ± 0.91	−11.06 ± 1.22 [a,b,c]	−10.49 ± 0.39 [a,b,c]
7	−8.51 ± 1.13	−7.12 ± 0.60	−9.31 ± 0.30	−11.88 ± 1.82 [a,b,c]	−14.58 ± 0.75 [a,b,c,d]
8	−8.89 ± 0.64	−8.15 ± 0.75	−10.24 ± 0.94	−14.34 ± 1.28 [a,b,c]	−14.54 ± 1.32 [a,b,c]
9	−10.15 ± 0.98	−8.43 ± 1.16	−10.43 ± 0.64	−15.42 ± 1.20 [a,b,c]	−16.12 ± 0.71 [a,b,c]
10	−9.81 ± 0.79	−9.41 ± 0.52	−11.88 ± 0.66 [a,b]	−16.72 ± 0.90 [a,b,c]	−16.96 ± 0.38 [a,b,c]
11	−10.97 ± 1.08	−12.84 ± 0.80	−13.76 ± 1.41 [a]	−17.34 ± 1.45 [a,b,c]	−16.58 ± 0.86 [a,b,c]
12	−10.86 ± 0.73	−14.74 ± 0.82 [a]	−17.26 ± 0.70 [a,b]	−18.84 ± 0.68 [a,b]	−17.34 ± 1.36 [a,b]

[a] Statistically significant differences vs. control group, $p < 0.05$. [b] Statistically significant differences vs. modified erythrocyte membranes with PS-NH$_2$-200 (C = 0.002 mg/mL), $p < 0.05$. [c] Statistically significant differences vs. modified erythrocyte membranes with PS-NH$_2$-200 (C = 0.01 mg/mL), $p < 0.05$. [d] Statistically significant differences vs. modified erythrocyte membranes with PS-NH$_2$-200 (C = 0.1 mg/mL), $p < 0.05$.

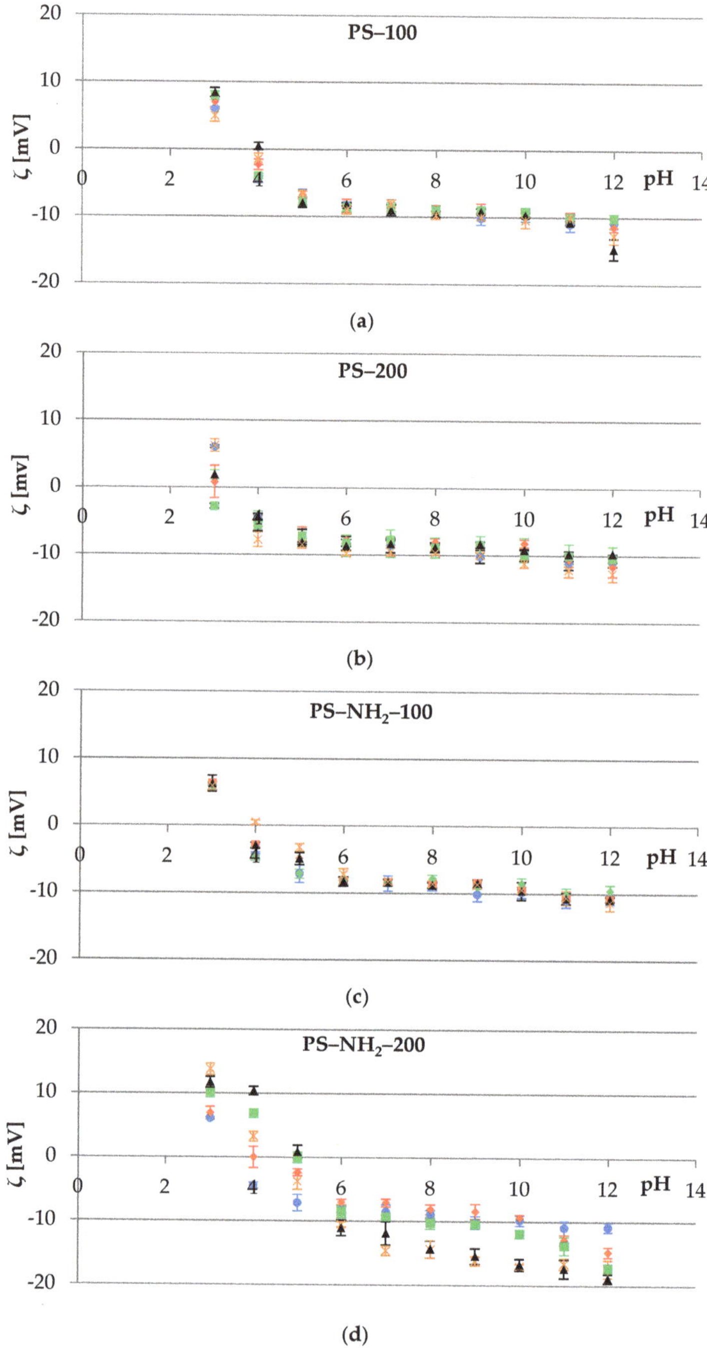

Figure 3. The zeta potential of the erythrocytes' cell membranes as a function of the pH of the electrolyte solution. Membranes untreated (•) or treated with 0.002 (♦), 0.01 (■), 0.1 (▲), and 0.5 (×) mg/mL of various types of polystyrene polymer: (**a**) PS–100, (**b**) PS–200, (**c**) PS–NH$_2$–100, and (**d**) PS–NH$_2$–200.

Erythrocytes, the same as most biological surfaces, can be characterized by the negative charge occurring on their surface. This attribute is due to the presence of sialic acid located on the glycoproteins on the exposed parts of the cell [42]. As it is rather troublesome to measure the true surface charge atop cells, zeta potential is used interchangeably, having been deduced from electrophoretic mobility measurements. This parameter is a crucial characteristic of biological membranes, as it is bound to stabilize red blood cells dispersed in an electrolyte solution as it repels erythrocytes from other types of cells and, especially, themselves. This way, adhesion between RBCs and interactions with the endothelium is regulated [43]. Durocher et al. [44] proved mature cells have less sialic acid and less surface charge than those still maturing. These changes were theorized to occur during the senescence.

Due to electrostatic interactions between negatively charged erythrocytes and ions of positive charge present in the structure of polystyrene, a disruption of the natural process occurs. Furthermore, while treating red blood cells with $PS-NH_2-200$, the zeta potential values change as a result of generating additional surface charge atop the cells. Based on the data presented in Figure 3d, it can be presumed that $PS-NH_2-200$ is attached to the RBC's membrane. As a consequence of this phenomenon, the zeta potential values of erythrocytes treated with this type of plastic are increased in the range of pH 2–7. However, ζ tends to decrease in a solution with a pH above 7 as OH^- ions start to react with the amine groups of the $PS-NH_2-200$ particles.

The process of adhesion to the membranes and internalization into the cells of many particles is influenced by their physical properties, including size, shape, solubility, surface composition, and surface charge [45,46]. This leads to polymers of identical structures but distinct diameters causing different, sometimes contrary, effects on the natural surfaces, as size is a determinant of the mechanism of interaction between plastic and biological membranes [47].

Functionalized polystyrene particles become adhered to erythrocytes due to distinct mechanisms, hydrogen bonding, hydrophobic interactions, and specific van der Waals forces, to name a few [48,49]. As has already been described, PS with carboxyl groups present on their surface ranging in size from 100 nm to 1.1 μm and non- and amine-functionalized PS 200 nm in diameter become easily attached to erythrocyte surfaces [48,49]. It was also proven that increasing the number of particles present per cell led to more particles attaching to the red blood cells' membranes. Erythrocytes' morphology did not change due to the adhesion, and the investigated process was observed to be non-specific to any area of the cell surfaces [48,49]. However, polystyrene submicron particles of no charge, as well as of positive and negative charges, were found inside the red blood cells (while they were smaller than 0.1 μm in diameter) in addition to being present on the cell surface (while ranging from 0.2 to 1 μm in size) [40]. What is more, PS 78 nm and 200 nm in diameter were found within erythrocytes without inhabiting RBC's membranes [41].

Platelets, similarly to erythrocytes, possess membrane glycoproteins that play an important role during two crucial processes: Adhesion to the subendothelial matrix and platelet–platelet cohesion, or aggregation [50]. These glycoproteins in their structure contain carboxyl groups that provide a negative charge to the surface of thrombocytes. The results for experimental studies of platelets' cell membranes, treated and untreated with polystyrene polymers, are shown in Figure 4. Slight ζ changes were observed for membranes in contact with PS-100, PS-200, and $PS-NH_2-100$ (Figure 4a–c), and the most visible were noted at extreme pH values, e.g., at pH~12 when 0.1 or 0.5 mg/mL PS-200 was used (Figure 4b). It was theorized that the changes result from the destruction of the cell membrane structure in an excessively alkaline environment.

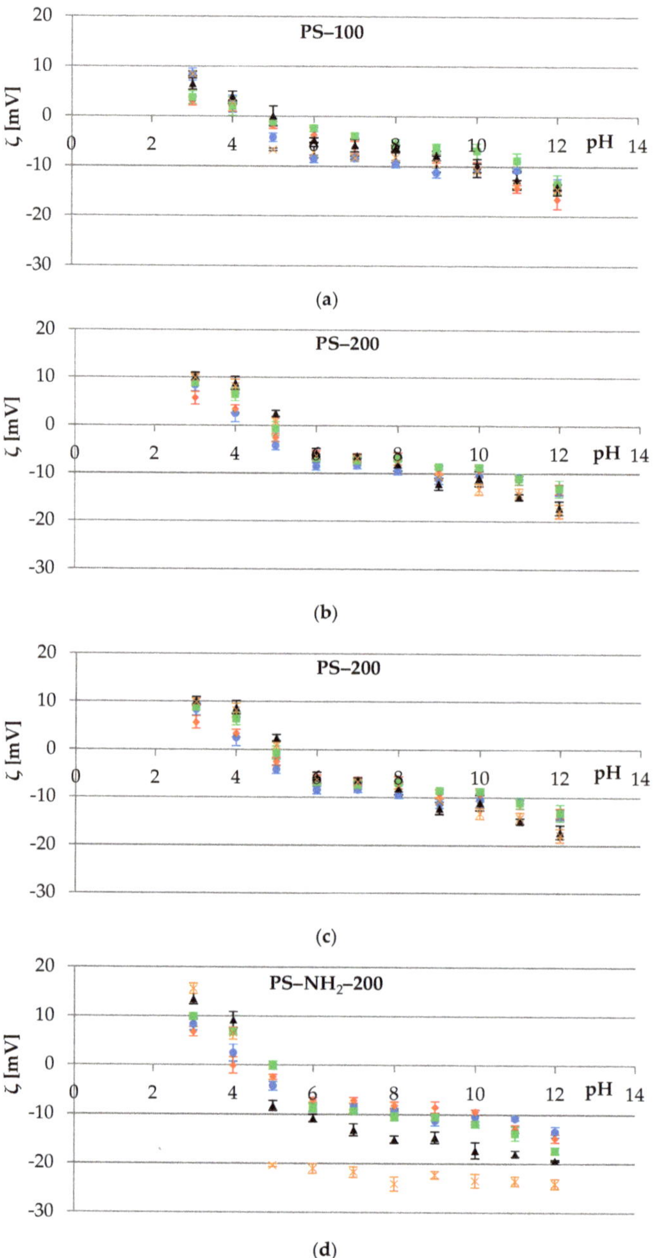

Figure 4. The zeta potential of the platelets' cell membranes as a function of the pH of the electrolyte solution. Membranes untreated (•) or treated with 0.002 (♦), 0.01 (■), 0.1 (▲), and 0.5 (×) mg/mL of various types of polystyrene polymer: (**a**) PS–100, (**b**) PS–200, (**c**) PS–NH$_2$–100, and (**d**) PS–NH$_2$–200.

Statistically significant changes, as in the case of erythrocytes, were observed for the thrombocytes' cell membranes treated with PS-NH$_2$-200 (Figure 4d), and the most visible in the entire pH range were noted for polymer concentrations of 0.1 and 0.5 mg/mL. At an acidic pH (pH~3 and pH~4), the ζ reached higher positive values, while in the remaining pH range (pH 5 to 12), the zeta potential increased towards negative values. The changes that occurred in the parameter with the participation of lower concentrations of polystyrene (0.002 and 0.01 mg/mL) were insignificant.

The data of the zeta potential of thrombocytes during PS-NH$_2$-200 treatment as a function of pH is compiled in Table 3.

Table 3. The zeta potential of the platelets' cell membranes after exposure to PS-NH$_2$-200 (C = 0.002, 0.01, 0.1, and 0.5 mg/mL).

pH	Control	0.002	0.01	0.1	0.5
3	8.40 ± 0.58	6.86 ± 0.98	9.97 ± 0.65	13.48 ± 1.07 [b]	15.28 ± 0.89 [a,b,c]
4	2.49 ± 0.77	-0.02 ± 1.65	6.85 ± 0.67 [a,b]	9.25 ± 1.62 [a,b]	7.17 ± 0.93 [a,b]
5	-4.19 ± 0.85	-2.37 ± 0.52	0.00 ± 0.83 [a,b]	-8.24 ± 1.00 [a,b,c]	-22.18 ± 1.82 [a,b,c,d]
6	-8.48 ± 1.72	-7.03 ± 0.45 [a]	-8.53 ± 0.91 [b]	-10.78 ± 0.76 [a,b,c]	-21.14 ± 0.95 [a,b,c,d]
7	-8.35 ± 1.24	-7.12 ± 0.60	-9.31 ± 0.30 [b]	-13.10 ± 1.21 [a,b,c]	-22.48 ± 1.61 [a,b,c,d]
8	-9.42 ± 0.74	-8.15 ± 0.75	-10.24 ± 0.94 [b]	-15.10 ± 0.72 [a,b,c]	-22.40 ± 0.90 [a,b,c,d]
9	-11.20 ± 0.97	-8.43 ± 1.16 [a]	-10.43 ± 0.64	-14.68 ± 1.16 [a,b,c]	-21.98 ± 1.54 [a,b,c,d]
10	-10.48 ± 1.49	-9.41 ± 0.52	-11.88 ± 0.66 [b]	-17.30 ± 1.65 [a,b,c]	-22.98 ± 0.78 [a,b,c,d]
11	-10.76 ± 0.38	-12.84 ± 0.80 [a]	-13.76 ± 1.41 [a]	-17.94 ± 0.61 [a,b,c]	-23.30 ± 1.03 [a,b,c,d]
12	-13.44 ± 1.08	-14.74 ± 0.82	-17.26 ± 0.70 [a,b]	-19.34 ± 0.15 [a,b,c]	-24.92 ± 1.16 [a,b,c,d]

[a] Statistically significant differences vs. control group, $p < 0.05$. [b] Statistically significant differences vs. modified erythrocyte membranes with PS-NH$_2$-200 (C = 0.002 mg/mL), $p < 0.05$. [c] Statistically significant differences vs. modified erythrocyte membranes with PS-NH$_2$-200 (C = 0.01 mg/mL), $p < 0.05$. [d] Statistically significant differences vs. modified erythrocyte membranes with PS-NH$_2$-200 (C = 0.1 mg/mL), $p < 0.05$.

There is much evidence in the literature that suggests the possibility of interactions between submicron polymer particles and elements/components of the circulatory system—blood and blood vessels. The existence of such interactions would explain the proatherothrombic effect observed in a number of models. Smyth et al. [51] conducted research investigating the ability of non-functionalized and functionalized polystyrene particles of different sizes (50 and 100 nm) to cause thrombocytes' aggregation in vitro and in vivo. Their results confirmed that the process of aggregation was influenced by the physical properties of PS. Tested particles caused GPIIb/IIIa-mediated aggregation of thrombocytes to a certain degree, varying due to the size and presence of surface groups. PS-NH$_2$ of 50 nm in diameter acted in an enhanced agonist-induced aggregation by linking adjacent platelets, many of which were not even activated. Nemmar et al. [52] incubated thrombocytes with 60 nm polystyrene particles with different surface chemistries and found that the potency of causing aggregation was highest when using aminated particles, followed by carboxylated ones. Unmodified polystyrene was observed to induce aggregation to the lowest degree. Although these findings were to be generally confirmed by McGuinnes et al. [53], a difference in causing aggregation between PS particles of amino and carboxyl groups present on their surfaces was not to be found.

Given the data acquired using the ELS technique presented in Figures 3 and 4 and Tables 2 and 3, it can be recognized that significant changes of zeta potential were observed when treating erythrocytes and thrombocytes' membranes using PS-NH$_2$-200. The exposition of cells' surfaces to 200 nm amine-functionalized polystyrene caused the electrokinetic potential to shift towards more negative values. As it was reported in the scientific literature, the larger the PS-NH$_2$ was, the higher ζ tended to increase [54], which was confirmed in Section 3.2.1. We hypothesized that the interaction between erythrocyte and thrombocyte membranes and the PS-NH$_2$-200 underlie the increased membrane perturbation. McGuinnes et al. [53] demonstrated that amino-functionalized PS appear to act via an unexplained mechanism that results in the display of anionic phospholipids on the outer surface of platelets and erythrocyte membranes and suggested that unusual protein adsorption patterns might lead to membrane perturbation-mediated aggregation.

3.2.2. The Effect of the Exposure Time

Subsequent studies were focused on the measurement of the zeta potential of red blood cell and platelet membranes as a function of pH depending on the exposure time of the membranes to different polymers. Experimental research was performed immediately after suspending the blood components in a 0.155 M NaCl solution containing the polymer at a concentration of 0.1 mg/mL (at which changes in relation to the control sample were observed) and after 1 h and 3 h exposure of erythrocytes and thrombocytes to the polymer.

The dependence of the zeta potential as a function of pH for RBCs is shown in Figure 5. Statistically significant changes were observed at the lowest and the highest pH values. Moreover, changes were observed in the case of (1) PS-100 at pH 4–5 (Figure 5a), (2) PS-NH$_2$-100 in a pH range from 4 to 6 (Figure 5c), and (3) PS-NH$_2$-200 at pH = 4 (Figure 5d).

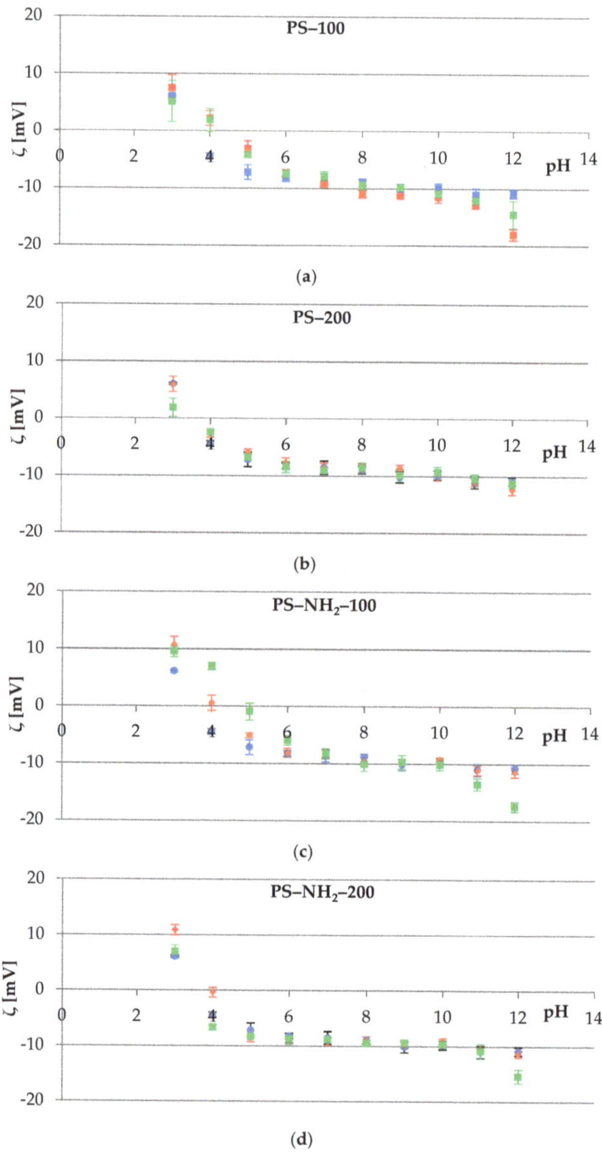

Figure 5. Zeta potential of the erythrocytes' cell membranes as a function of the pH of the electrolyte solution. The parameter was measured immediately (•), after 1 h (♦), and 3 h (■) of exposing the sample to the polymer: (**a**) PS–100, (**b**) PS–200, (**c**) PS–NH$_2$–100, and (**d**) PS–NH$_2$–200.

All particles used in this study were associated with erythrocytes—attached to or internalized by cells, depending on the size and surface chemistry. From the data shown in Figure 5, it can be seen that both processes occurred quickly and efficiently within one hour of the cells and particles coming into contact. Any coincidental discrepancies between the zeta values obtained for different durations of cell exposure to the polymer are likely due to the possibility of particles detaching from erythrocytes, as a result of shear forces, cell–cell interactions, and cell–vessel wall interactions [49].

In the case of studies on thrombocyte membranes (Figure 6), significant changes in the ζ potential were observed while exposing these cells to PS-100 and PS-NH$_2$-200.

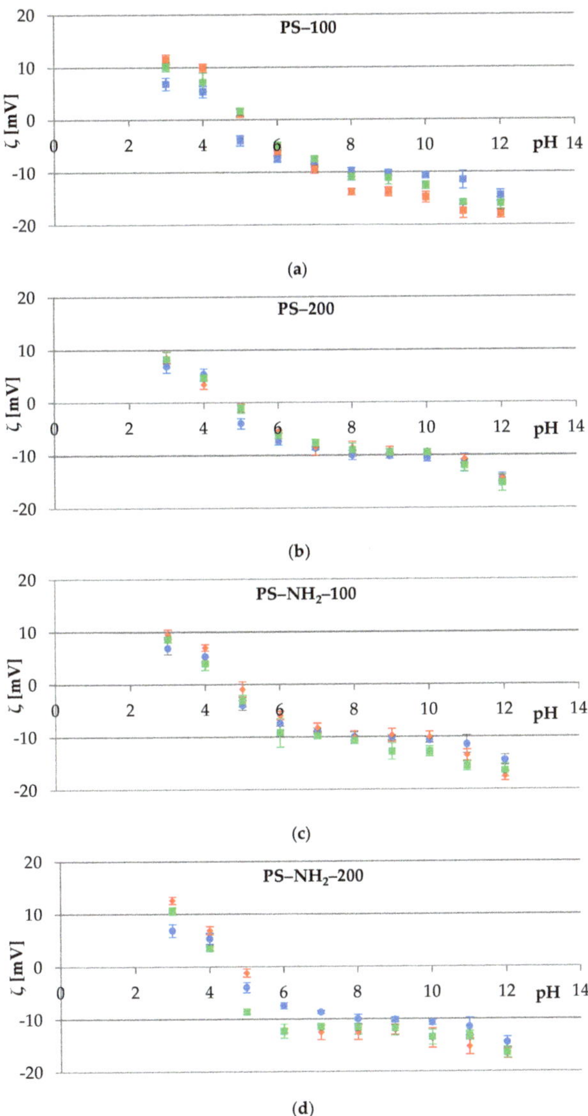

Figure 6. Zeta potential of the platelets' cell membranes as a function of the pH of the electrolyte solution. The parameter was measured immediately (●), after 1 h (♦), and 3 h (■) of exposing the sample to the polymer: (**a**) PS–100, (**b**) PS–200, (**c**) PS–NH$_2$–100, and (**d**) PS–NH$_2$–200.

From the data depicted in Figure 6, it can be noted that PS-200 and PS-NH$_2$-100 quickly and efficiently attached to platelet membranes within an hour. For PS-100 and PS-NH$_2$-200 particles, the association of polystyrene with cells was slower. This can be seen in the obtained dependencies, in which the zeta potential values measured after 3 h of thrombocyte exposure to the polymer are more similar to those measured directly than to those determined after 1 h of cell exposure. In the subject literature, studies have been presented in which internalization was noticeable as early as 1 min after incubation [46], and in which, as the duration of cell exposure to polystyrene increased, a greater probability of polystyrene association with cells was reported [37].

4. Conclusions

To summarize, there is a need to understand the influence of submicron polystyrene particle properties, such as size and chemical composition, on the various endpoints of particle toxicity. In the presented research, we investigated whether polystyrene particles influence the electrokinetic potential of erythrocytes and thrombocytes. Two types of polymers were used, and each experiment was conducted several times. In addition, different particle sizes and concentrations, as well as exposure durations to the polymer, always provided the same results, i.e., the internalization of PS-100, PS-200, and PS-NH$_2$-100 particles into the cells and the attachment of PS-NH$_2$-200 particles to cell membranes. Our study clearly shows that the size and surface chemistry of PS particles determine the extent to which they affect the electrical properties of thrombocyte and erythrocyte membranes.

Author Contributions: Conceptualization, M.Z. and M.N.; methodology, M.Z., J.K., J.B. and M.N.; validation, M.Z., J.K., J.B. and M.N.; formal analysis, M.Z., J.K. and M.N.; investigation, M.Z.; resources, M.Z., J.K. and M.N.; data curation, J.K. and M.N.; writing—original draft preparation, M.Z., J.K. and M.N.; writing—review and editing, M.Z., J.K., M.W. and M.N.; visualization, M.Z., J.K. and M.N.; supervision, M.N. All authors have read and agreed to the published version of the manuscript.

Funding: The FTIR spectrometer was funded by the EU as part of the Operational Program Development of Eastern Poland 2007-2013 (projects Nr POPW.01.03.00-20-034/09-00 and POPW.01.03.00-20-004/11-00, the Centre of Synthesis and Analysis BioNanoTechno, University of Bialystok, Poland).

Institutional Review Board Statement: Not applicable.

Informed Consent Statement: Not applicable.

Data Availability Statement: Not applicable.

Conflicts of Interest: The authors declare no conflict of interest.

Abbreviations

ELS	Electrophoretic light scattering
DLS	Dynamic light scattering
FTIR	Fourier transform infrared
ζ	Zeta potential (electrokinetic potential)
PDI	Polydispersity index
PS	Polystyrene particles
PS-100	100 nm polystyrene particles
PS-200	200 nm polystyrene particles
PS-NH$_2$-100	100 nm polystyrene particles with amino groups
PS-NH$_2$-200	200 nm polystyrene particles with amino groups
RBC	Red blood cells (erythrocytes)
Platelets (thrombocytes)	

References

1. Kazemi, M.; Fini, E.H. State of the art in the application of functionalized waste polymers in the built environment. *Resour. Conserv. Recycl.* **2022**, *177*, 105967. [CrossRef]
2. Lai, J.H. *Polymers for Electronic Applications*, 1st ed.; CRC Press: Boca Raton, FL, USA, 2018.
3. Puoci, F. *Advanced Polymers in Medicine*, 1st ed.; Springer: Cham, Switzerland, 2015.
4. Manivannan, B.; Eltzov, E.; Borisover, M. Submicron polymer particles may mask the presence of toxicants in wastewater effluents probed by reporter gene containing bacteria. *Sci. Rep.* **2021**, *11*, 7424. [CrossRef] [PubMed]
5. Kik, K.; Bukowska, B.; Sicińska, P. Polystyrene Nanoparticles: Sources, Occurrence in the Environment, Distribution in Tissues, Accumulation and Toxicity to Various Organisms. *Environ. Pollut.* **2020**, *262*, 114297. [CrossRef] [PubMed]
6. Zimmermann, W. Biocatalytic recycling of polyethylene terephthalate plastic. *Philos. Trans. A Math. Phys. Eng. Sci.* **2020**, *378*, 20190273. [CrossRef]
7. Fotopoulou, K.N.; Karapanagioti, H.K. Degradation of Various Plastics in the Environment. In *Hazardous Chemicals Associated with Plastics in the Marine Environment*; Takada, H., Karapanagioti, H., Eds.; The Handbook of Environmental Chemistry Series; Springer: Cham, Switzerland, 2017; Volume 78.
8. Roohi; Bano, K.; Kuddus, M.; Zaheer, M.R.; Zia, Q.; Khan, M.F.; Ashraf, G.M.; Gupta, A.; Aliev, G. Microbial enzymatic degradation of biodegradable plastics. *Curr. Pharm. Biotechnol.* **2017**, *18*, 429–440. [CrossRef]
9. Urbanek, A.K.; Rymowicz, W.; Mirończuk, A.M. Degradation of plastics and plastic-degrading bacteria in cold marine habitats. *Appl. Microbiol. Biotechnol.* **2018**, *102*, 7669–7678. [CrossRef]
10. Yang, S.-S.; Brandon, A.M.; Andrew Flanagan, J.C.; Yang, J.; Ning, D.; Cai, S.-Y.; Fan, H.-Q.; Wang, Z.-Y.; Ren, J.; Benbow, E. Biodegradation of polystyrene wastes in yellow mealworms. *Chemosphere* **2018**, *191*, 979–989. [CrossRef]
11. Iqbal, S.; Xu, J.; Allen, S.D.; Khan, S.; Nadir, S.; Arif, M.S.; Yasmeen, T. Unraveling consequences of soil micro- and nano-plastic pollution on soil-plant system: Implications for nitrogen (N) cycling and soil microbial activity. *Chemosphere* **2020**, *260*, 127578. [CrossRef]
12. Blettler, M.C.M.; Abrial, E.; Khan, F.R.; Sivri, N.; Espinola, L.A. Freshwater plastic pollution: Recognizing research biases and identifying knowledge gaps. *Water Res.* **2018**, *143*, 416–424. [CrossRef]
13. Haward, M. Plastic pollution of the world's seas and oceans as a contemporary challenge in ocean governance. *Nat. Commun.* **2018**, *9*, 667. [CrossRef]
14. Cook, C.R.; Halden, R.U. Ecological and health issues of plastic waste. In *Plastic Waste and Recycling*; Letcher, T.M., Ed.; Academic Press: Cambridge, MA, USA, 2020; pp. 513–527.
15. Frydkjær, C.K.; Iversen, N.; Roslev, P. Ingestion and egestion of microplastics by the cladoceran daphnia magna: Effects of regular and irregular shaped plastic and sorbed phenanthrene. *Bull. Environ. Contam. Toxicol.* **2017**, *99*, 655–661. [CrossRef] [PubMed]
16. Muradov, M.; Kot, P.; Ateeq, M.; Abdullah, B.; Shaw, A.; Hashim, K.; Al-Shamma'a, A. Real-time detection of plastic shards in cheese using microwave-sensing technique. *Proceedings* **2019**, *42*, 54.
17. Rehse, S.; Kloas, W.; Zarfl, C. Short-term exposure with high concentrations of pristine microplastic particles leads to immobilisation of daphnia magna. *Chemosphere* **2016**, *153*, 91–99. [CrossRef] [PubMed]
18. Jemec, A.; Horvat, P.; Kunej, U.; Bele, M.; Kržan, A. Uptake and effects of microplastic textile fibers on freshwater crustacean daphnia magna. *Environ. Pollut.* **2016**, *219*, 201–209. [CrossRef] [PubMed]
19. Rothschild, M.F.; Ruvinsky, A. *The Genetics of the Pig*, 2nd ed.; C.A.B. International: Wallingford, UK, 2011.
20. Vodička, P.; Smetana, K.; Dvořánková, B.; Emerick, T.; Xu, Y.Z.; Ourednik, J.; Ourednik, V.; Motlík, J. The miniature pig as an animal model in biomedical research. *Ann. N. Y. Acad. Sci.* **2005**, *1049*, 161–171. [CrossRef]
21. Swindle, M.M.; Makin, A.; Herron, A.J.; Clubb, F.J.; Frazier, K.S. Swine as models in biomedical research and toxicology testing. *Vet. Pathol.* **2012**, *49*, 344–356. [CrossRef]
22. Aigner, B.; Renner, S.; Kessler, B.; Klymiuk, N.; Kurome, M.; Wünsch, A.; Wolf, E. Transgenic pigs as models for translational biomedical research. *J. Mol. Med.* **2010**, *88*, 653–664. [CrossRef]
23. Bradbury, A.G.; Argyle, S.; Eddleston, M.; Clutton, R.E. Prophylactic use of antimicrobials in surgical pig models; a literature review. *Vet. Res.* **2015**, *177*, 16–21.
24. Campisi, C.C.; Jiga, L.P.; Ryan, M.; di Summa, P.G.; Campisi, C.; Ionac, M. Mastering lymphatic microsurgery: Anew training model in living tissue. *Ann. Plast. Surg.* **2017**, *79*, 298–303. [CrossRef]
25. Henze, L.J.; Koehl, N.J.; O'Shea, J.P.; Kostewicz, E.S.; Holm, R.; Griffin, B.T. The pig as a preclinical model for predicting oral bioavailability and in vivo performance of pharmaceutical oral dosage forms: A pearrl review. *J. Pharm. Pharmacol.* **2019**, *71*, 581–602. [CrossRef]
26. Bures, J.; Kvetina, J.; Radochova, V.; Tacheci, I.; Peterova, E.; Herman, D.; Dolezal, R.; Kopacova, M.; Rejchrt, S.; Douda, T. The pharmacokinetic parameters and the effect of a single and repeated doses of memantine on gastric myoelectric activity in experimental pigs. *PLoS ONE* **2020**, *15*, e0227781. [CrossRef]
27. Peng, L.; Fu, D.; Qi, H.; Lan, C.Q.; Yu, H.; Ge, C. Micro- and nano-plastics in marine environment: Source, distribution and threats—A review. *Sci. Total Environ.* **2020**, *698*, 13425. [CrossRef] [PubMed]
28. Wu, X.; Lu, J.; Du, M.; Xu, X.; Beiyuan, J.; Sarkar, B.; Bolan, N.; Xu, W.; Xu, S.; Chen, X.; et al. Particulate plastics-plant interaction in soil and its implications: A review. *Sci. Total Environ.* **2021**, *792*, 148337. [CrossRef] [PubMed]

29. Beriot, N.; Zomer, P.; Zornoza, R.; Geissen, V. A laboratory comparison of thei interactions between three plastic mulch types and 38 active substances found in pesticides. *PeerJ* **2020**, *8*, e9876. [CrossRef] [PubMed]

30. Clukey, K.E.; Lepczyk, C.A.; Balazs, G.H.; Work, T.M.; Lynch, J.M. Investigation of plastic debris ingestion by four species of sea turtles collected as bycatch in pelagic pacific longline fisheries. *Mar. Pollut. Bull.* **2017**, *120*, 117–125. [CrossRef] [PubMed]

31. Markic, A.; Gaertner, J.-C.; Gaertner-Mazouni, N.; Koelmans, A.A. Plastic ingestion by marine fish in the wild. *Crit. Rev. Environ. Sci. Technol.* **2020**, *50*, 657–697. [CrossRef]

32. Choi, D.; Hwang, J.; Bang, J.; Han, S.; Kim, T.; Oh, Y.; Hwang, Y.; Choi, J.; Hong, J. In vitro toxicity from a physical perspective of polyethylene microplastics based on statistical curvature change analysis. *Sci. Total Environ.* **2021**, *752*, 142242. [CrossRef]

33. Marsalek, R.; Kotyrba, M.; Volna, E.; Jarusek, R. Neural network modelling for prediction of zeta potential. *Mathematics* **2021**, *9*, 3089. [CrossRef]

34. Naumowicz, M.; Zając, M.; Kusaczuk, M.; Gál, M.; Kotyńska, J. Electrophoretic light scattering and electrochemical impedance spectroscopy studies of lipid bilayers modified by cinnamic acid and its hydroxyl derivatives. *Membranes* **2020**, *10*, 343. [CrossRef]

35. Wang, M.L.; Hauschka, P.V.; Tuan, R.S.; Steinbeck, M.J. Exposure to Particles Stimulates Superoxide Production by Human THP-1 Macrophages and Avian HD-11EM Osteoclasts Activated by Tumor Necrosis Factor-α and PMA. *J. Arthroplast.* **2002**, *17*, 335–346. [CrossRef]

36. Liu, S.; Wu, X.; Gu, W.; Yu, J.; Wu, B. Influence of the digestive process on intestinal toxicity of polystyrene microplastics as determined by in vitro Caco-2 models. *Chemosphere* **2020**, *256*, 127204. [CrossRef] [PubMed]

37. Wu, B.; Wu, X.; Liu, S.; Wang, Z.; Chen, L. Size-dependent effects of polystyrene microplastics on cytotoxicity and efflux pump inhibition in human Caco-2 cells. *Chemosphere* **2019**, *221*, 333–341. [CrossRef] [PubMed]

38. Fang, J.; Xuan, Y.; Li, Q. Preparation of polystyrene spheres in different particle sizes and assembly of the PS colloidal crystals. *Sci. China Technol. Sci.* **2010**, *53*, 3088–3093. [CrossRef]

39. Joseph, E.; Singhvi, G. Multifunctional nanocrystals for cancer therapy: A potential nanocarrier. In *Nanomaterials for Drug Delivery and Therapy*; Grumezescu, A.M., Ed.; William Andrew Publishing: Amsterdam, The Netherlands, 2019; pp. 91–116.

40. Silva, D.C.N.; Jovino, C.N.; Silva, C.A.L.; Fernandes, H.P.; Filho, M.M.; Lucena, S.C.; Costa, A.M.D.N.; Cesar, C.L.; Barjas-Castro, M.L.; Santos, B.S. Optical tweezers as a new biomedical tool to measure zeta potential of stored red blood cells. *PLoS ONE* **2012**, *7*, e31778. [CrossRef]

41. Godin, C.; Caprani, A. Effect of blood storage on erythrocyte/wall interactions: Implications for surface charge and rigidity. *Eur. Biophys. J.* **1997**, *26*, 175–182. [CrossRef]

42. Durocher, J.R.; Payne, R.C.; Conrad, M.E. Role of sialic acid in erythrocyte survival. *Blood* **1975**, *45*, 11–20. [CrossRef]

43. Wang, T.; Bai, J.; Jiang, X.; Nienhaus, G.U. Cellular uptake of nanoparticles by membrane penetration: Astudy combining confocal microscopy with FTIR spectroelectrochemistry. *ACS Nano* **2012**, *6*, 1251–1259. [CrossRef]

44. Jiang, X.; Dausend, J.; Hafner, M.; Musyanovych, A.; Röcker, C.; Landfester, K.; Mailänder, V.; Nienhaus, G.U. Specific effects of surface amines on polystyrene nanoparticles in their interactions with mesenchymal stem cells. *Biomacromolecules* **2010**, *11*, 748–753. [CrossRef]

45. Dwivedi, M.V.; Harishchandra, R.K.; Koshkina, O.; Maskos, M.; Galla, H.-J. Size influences the effect of hydrophobic nanoparticles on lung surfactant model systems. *Biophys. J.* **2014**, *106*, 289–298. [CrossRef]

46. Chambers, E.; Mitragotri, S. Prolonged circulation of large polymeric nanoparticles by non-covalent adsorption on erythrocytes. *J. Control. Release* **2004**, *100*, 111–119. [CrossRef]

47. Chambers, E.; Mitragotri, S. Long circulating nanoparticles via adhesion on red blood cells: Mechanism and extended circulation. *Exp. Biol. Med.* **2007**, *232*, 958–966.

48. Rothen-Rutishauser, B.M.; Schürch, S.; Haenni, B.; Kapp, N.; Gehr, P. Interaction of fine particles and nanoparticles with red blood cells visualized with advanced microscopic techniques. *Environ. Sci. Technol.* **2006**, *40*, 4353–4359. [CrossRef] [PubMed]

49. Geiser, M.; Rothen-Rutishauser, B.; Kapp, N.; Schürch, S.; Kreyling, W.; Schulz, H.; Semmler, M.; Im Hof, V.; Heyder, J.; Gehr, P. Ultrafine particles cross cellular membranes by nonphagocytic mechanisms in lungs and in cultured cells. *Environ. Health Perspect.* **2005**, *113*, 1555–1560. [CrossRef] [PubMed]

50. Kunicki, T.J. Platelet membrane glycoproteins and their function: An overview. *Blut* **1989**, *59*, 30–34. [CrossRef]

51. Smyth, E.; Solomon, A.; Vydyanath, A.; Luther, P.K.; Pitchford, S.; Tetley, T.D.; Emerson, M. Induction and enhancement of platelet aggregation in vitro and in vivo by model polystyrene nanoparticles. *Nanotoxicology* **2015**, *9*, 356–364. [CrossRef]

52. Nemmar, A.; Hoylaerts, M.F.; Hoet, P.H.M.; Dinsdale, D.; Smith, T.; Xu, H.; Vermylen, J.; Nemery, B. Ultrafine Particles affect experimental thrombosis in an in vivo hamster model. *Am. J. Respir. Crit. Care Med.* **2002**, *166*, 998–1004. [CrossRef]

53. McGuinnes, C.; Duffin, R.; Brown, S.; Mills, L.N.; Megson, I.L.; MacNee, W.; Johnston, S.; Lu, S.L.; Tran, L.; Li, R.; et al. Surface derivatization state of polystyrene latex nanoparticles determines both their potency and their mechanism of causing human platelet aggregation in vitro. *Toxicol. Sci.* **2011**, *119*, 359–368. [CrossRef]

54. Oslakovic, C.; Cedervall, T.; Linse, S.; Dahlbäck, B. Polystyrene nanoparticles affecting blood coagulation. *Nanomedicine* **2012**, *8*, 981–986. [CrossRef]

MDPI

St. Alban-Anlage 66

4052 Basel

Switzerland

Tel. +41 61 683 77 34

Fax +41 61 302 89 18

www.mdpi.com

Membranes Editorial Office

E-mail: membranes@mdpi.com

www.mdpi.com/journal/membranes